Topics in Mining, Metallurgy and Materials Engineering

Series editor

Carlos P. Bergmann, Porto Alegre, Brazil

"Topics in Mining, Metallurgy and Materials Engineering" welcomes manuscripts in these three main focus areas: Extractive Metallurgy/Mineral Technology; Manufacturing Processes, and Materials Science and Technology. Manuscripts should present scientific solutions for technological problems. The three focus areas have a vertically lined multidisciplinarity, starting from mineral assets, their extraction and processing, their transformation into materials useful for the society, and their interaction with the environment.

More information about this series at http://www.springer.com/series/11054

Hossein Aghajani · Sahand Behrangi

Plasma Nitriding of Steels

Springer

Hossein Aghajani
Department of Materials Engineering
University of Tabriz
Tabriz
Iran

Sahand Behrangi
Department of Materials Engineering
University of Tabriz
Tabriz
Iran

ISSN 2364-3293 ISSN 2364-3307 (electronic)
Topics in Mining, Metallurgy and Materials Engineering
ISBN 978-3-319-82730-8 ISBN 978-3-319-43068-3 (eBook)
DOI 10.1007/978-3-319-43068-3

Printed on acid-free paper

This Springer imprint is published by Springer Nature
The registered company is Springer International Publishing AG
The registered company address is: Gewerbestrasse 11, 6330 Cham, Switzerland

Preface

Nowadays plasma nitriding is widely used in different industries and the researches and papers published in this field of surface engineering are growing every day to identify the role of influencing parameters on the process. The goal of all these studies is to optimize the parameters and achieve the best microstructural, mechanical, and corrosion properties. However, for an engineer (for example) there is no collective book containing briefly the results of the studies. Besides, it will be time-wasting to search and read all the recent researches and findings and summarize them to obtain the applicable ranges for process parameters. This book may partially act this role for readers. In this book, the results obtained by the researches are collected, summarized, and categorized. Therefore, each reader can easily find the desirable ranges for parameters to be applied. This will save the required time for multiple trial and errors.

The authors hope that this book be useful for readers and inspire other authors and researchers to write such books.

Tabriz, Iran

Hossein Aghajani
Sahand Behrangi

Contents

Chapter 1
Introduction

One of the most important applications of surface engineering is producing a hard layer on the surface of a tough material to improve, e.g., wear resistance. Therefore, different types of thermochemical treatments are used to obtain good surface properties. Among these methods, nitriding is extensively used in industry. The base of this method is diffusion of atomic nitrogen into the surface of steel while it is in ferritic state. This happens when holding the sample at 590 °C and lower (usually between 510 and 590 °C) [1].

Mechanism of nitriding refers to affinity of nitrogen for iron. At elevated temperatures, the BCC lattice of ferritic iron begins to vibrate. Due to this vibration, the small nitrogen atoms can diffuse into the iron lattice and form nitrides with iron or other alloying elements such as Al, Cr, Mo, V, and Ti. The formation of nitrides of these elements such as Cr_2N, Mo_2N, AlN, VN, and W_2N leads to the interlock of slip planes and consequent hardening of case [2, 3].

1.1 Surface Layers Formed by Nitriding

In general, the nitrogen-treated surface consists of two following regions:

1.1.1 Compound (White) Layer

The thickness of this top layer strongly depends on the chemical composition, e.g., in the case of plain carbon steel the compound layer is thicker compared to alloy steels. The compound layer mainly constituted by two iron nitrides of γ'-Fe_4N and ε-$Fe_{2-3}N$.

H. Aghajani and S. Behrangi, *Plasma Nitriding of Steels*, Topics in Mining, Metallurgy and Materials Engineering, DOI 10.1007/978-3-319-43068-3_1

In the surface where the nitrogen content is higher, the formation of ε-$Fe_{2-3}N$ is accelerated. However, with increasing the distance from surface toward core, the ε-$Fe_{2-3}N$ phase is replaced by γ′-Fe_4N and its amount decreases gradually.

A compound layer of predominately ε phase will create a surface with good wear characteristics, but it will have no impact strength. The presence of γ′-Fe_4N can improve the impact strength of the layer. Due to its brittle nature, in many applications no or a trace thickness of this layer is allowed.

1.1.2 Diffusion Layer

This layer which is formed under the compound layer is mainly composed of nitrides of alloying elements. In contrast to compound layer, the role of diffusion layer is crucial in increasing the case hardness and fatigue life of the substrate [1, 3, 4].

The enhanced hardness of nitrided steels is the result of the presence and distribution of the nitrides of alloying elements. In other words, with formation of compound layer, nitrogen diffuses inward and combines with nitride-forming elements to form nitride compounds. The precipitation of these nitrides in grain boundaries and specified crystallographic planes results in an increase in the hardness of diffusion layer [4].

Generally speaking, nitriding is applicable to wide variety of steels, e.g., carbon steels, low alloy steels, tool steels, and stainless steels. However, the existence of nitride-forming elements is essential for high efficiency of nitriding. The strong nitride-forming elements which form stable nitrides are Al, Cr, V, W, and Mo. An important limitation also exists about the amount of carbon in the steel to be nitrided that should not exceed 0.5 %. This is due to the increase in the possibility of alloying elements to form carbides and lowering the formation of nitrides [1, 2, 5].

1.2 Methods of Nitriding

1.2.1 Liquid Nitriding

This process takes place in a molten salt bath containing either cyanides or cyanates with temperature of 510–580 °C.

Because of low applied temperature, the diffusion of nitrogen is faster than carbon, so the carburizing did not happen. The composition of this bath could be either cyanide or cyanate. The most general composition is 60–70 % sodium salts and 30–40 % potassium salts. The salt bath should be aged before using. This action results in a reduction in the amount of cyanide and carbonate. The nitriding treatment usually done in the temperature range of 540–595 °C and the properties of nitride layer depend on the cyanide to cyanate ratio and the amount of existing cyanate in the bath [6].

1.2.2 *Gas Nitriding*

In this method, the nitriding temperature is between 500–565 °C and below Ac_1 (the austenite formation temperature for ferritic steel). Before reaching temperature to 150 °C, the existing air in the furnace should be evacuated. This action protects specimens from oxidation. The dry ammonia gas is blown into the furnace and decomposed on the surface of specimen. Thereafter, as a result of reactions with alloying elements, different nitrides are formed. Nitrogen in its atomic form is adsorbed by the steel surface [5, 6].

Gas nitriding is performed with two methods.

1.2.2.1 Single-Stage Method

In this procedure, the temperature is at 500–530 °C and dissociation of ammonia is between 15 and 30 %. The thickness of the white layer is almost 0.05 mm which results in surface embrittlement.

1.2.2.2 Two-Stage Method

Or "Floe process" is carried out at about 500–510 °C for the first 15–20 h. Thereafter the temperature is raised to 550 °C for the remaining of cycle. The degree of dissociation of ammonia is between 65 and 85 %. An important advantage of this process is lowering the compound layer thickness [2, 4].

1.2.3 *Plasma Nitriding*

Plasma is the fourth state of matter and extensively used for a large number of surface treatments such as spraying, vapor deposition, nitriding, etc. Plasma is formed by ionization of gas atoms by heat, radiation, or electrical discharge. In fact plasma is a conducting gas consisting of appreciable proportion of charged particles [7].

In plasma nitriding, elemental nitrogen is introduced to the surfaces of work pieces for subsequent diffusion into the metal. High-voltage electrical energy forms a plasma through which nitrogen atoms are accelerated to impinge on workpiece. Ion bombardment heats workpieces, cleans surfaces, and provides active nitrogen [6].

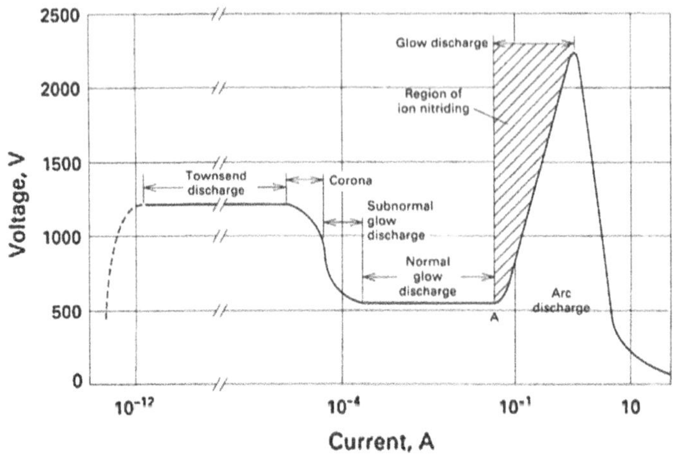

Fig. 1.1 The plot of voltage versus current density for formation of discharges in argon [8]

With applying lower voltages, a weak electrical current is produced. This initial current is almost constant. With increasing the voltage beyond a specified value, the electrical current causes electrons from the gas atoms within the vacuum chamber to leave the outer shell of electrons circulating around the nucleus of the atom. These released electrons are accelerated toward the counter electrode and impact other electrons and cause them to be accelerated in a chain-like manner. Moreover, some energetic electrons bombard the anode surface and make the electrons leave the surface and collide with others. As shown in Fig. 1.1, a sudden rise in current density occurs in this region called "Townsend discharge." The appropriate value of voltage and current density for plasma nitriding called "Glow discharge" is shown in Fig. 1.1. For this purpose, a uniform glow should be formed that will follow the shape and geometrical form of the workpiece. Beyond this region, arc discharging may occur and leads to burning and damaging of the surface. Therefore, the applying voltage for plasma nitriding must be chosen from this region [3]. To obtain this glow discharge in the nitriding process, the components to be treated are placed in the discharge vessel and cathodically charged to a potential of 400–1000 V, with respect to the wall of the vessel which acts as the anode of the circuit [2].

1.2.3.1 Mechanism of Plasma Nitriding

The most acceptable mechanism for case hardening by plasma nitriding is as follows. First of all, the interaction of atomic nitrogen with surface results in the formation of FeN. This nitride phase is unstable and is transformed to ε and γ' nitrides; therefore, the compound layer is formed. Moreover, with diffusing of nitrogen toward the core, the diffusion layer is formed. This is shown schematically

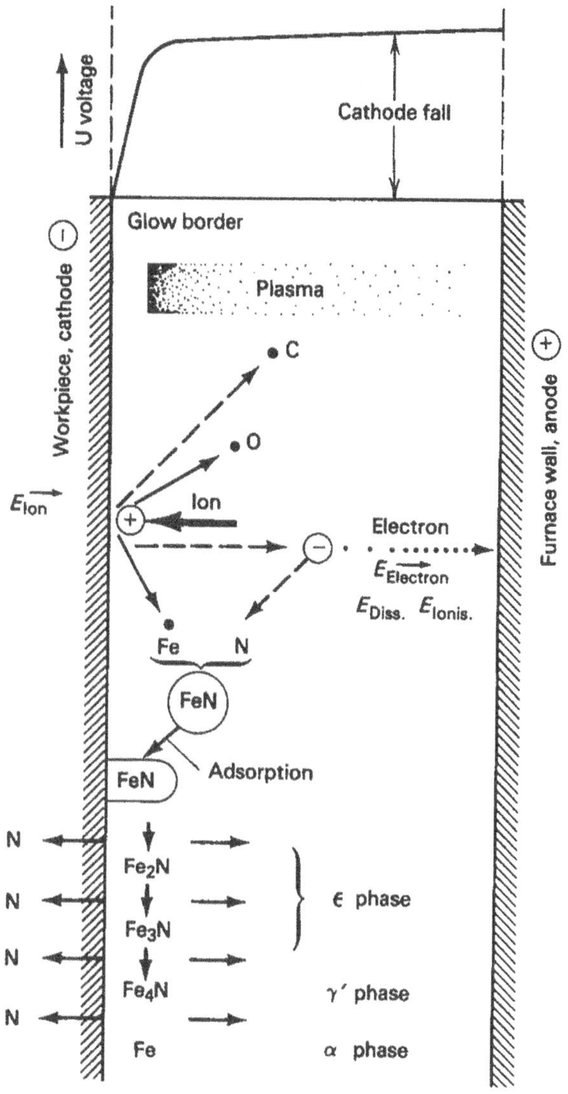

Fig. 1.2 The proposed mechanism for plasma nitriding [8]

in Fig. 1.2 [8]. Plasma process allows both bulk and grain-boundary diffusion compared to only grain-boundary diffusion in gas nitriding [7].

1.2.3.2 Parameters

There are several parameters influencing plasma nitriding process some of which are noted in the following.

Gas Mixture

The usual gas mixture used for plasma nitriding is 25 % N_2 + 75 % H_2. If the nitrogen amount raises to 50 %, the growth rate of case is increased. Further increment in nitrogen amount results in the thickening of compound layer [2]. It should be noted that, there are contrary opinions about the role of hydrogen in the nitriding process. Some researchers mentioned its positive role such as increasing reaction rate, thickness of nitrided case, and density of active nitriding species in the plasma. On the other hand, some researchers have concluded that hydrogen has no or deleterious effect [9, 10].

Temperature

The temperature for plasma nitriding plays a key role in nitriding depth and should be in the range of 350–580 °C. For alloyed steels, the required temperature is lower than carbon steels.

Time, pressure, and plasma power are other parameters influencing the condition of plasma nitriding [2, 11].

1.2.3.3 Parts of a Plasma Nitriding System

As shown schematically in Fig. 1.3, the main parts of a plasma nitriding system are a vacuum chamber, a power supply, and a process gas system with a gas-mixing panel or other mass flow controls. It should be noted that the work support fixturing should be isolated to warranty an electrical isolation between the workpiece and vacuum vessel.

The vacuum chamber can be a hot-walled design or a dual-walled and water-cooled design with no internal insulation because of low temperature needed for plasma nitriding and low risk of heat loss.

The power supply as the most important part of the system must provide an output voltage of 0–1000 V and an output current depending on the size of the vessel and workload (typically 25–450 A for DC current).

Another important design consideration in the power supply is arc detection/suppression controls. Arcing can occur because the glow discharge process causes the removal of surface impurities. This is done in the form of an arc, in which a sudden decrease in voltage and increase in current is occurred. When an arc is detected, the power output is shut off and the existing power diverted from the workload to avoid any possible damage.

The gas-mixing panel is used for mixing gases in exact ratios to obtain the specified microstructure.

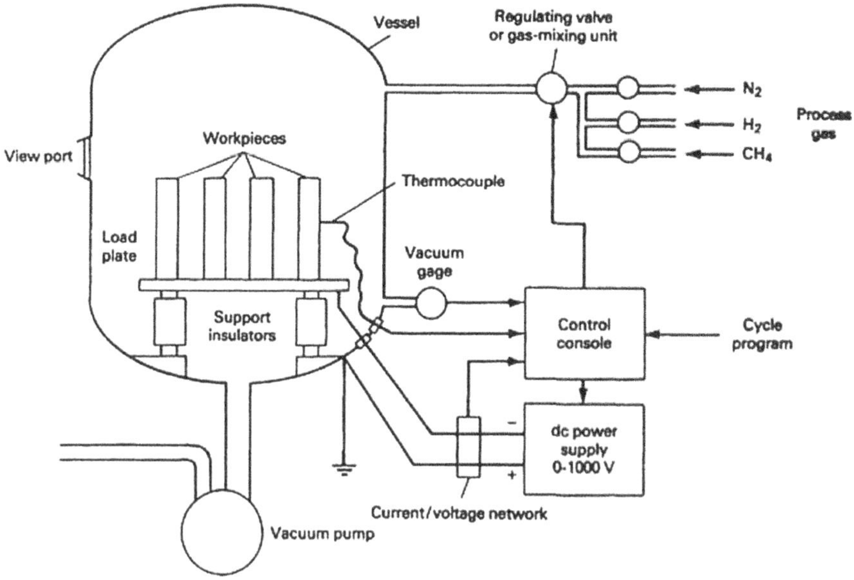

Fig. 1.3 Parts of a typical plasma nitriding system [8]

Plasma nitriding is generally performed at absolute pressure levels of 130–1300 Pa (1–10 torr) and therefore, an exact pressure control system is vital.

An auxiliary heating system is needed to compensate the insufficiency of glow discharge for heating the large workpieces [8].

1.2.3.4 Advantages and Disadvantages of Plasma Nitriding

Plasma nitriding has some advantages compared to other methods. Some of them are: (1) Lowering the process time, (2) Lowering distortion and dimensional changes, (3) No need for final finishing processes, (4) Lowering the finished cost of process. (5) The surface homogeneity and uniformity of samples with complex geometry, (6) Harmlessness of plasma nitriding, (7) Fast kinetics of reaction at the surface and (8) Possibility to remove the oxide films before nitriding to accelerate nitrogen mass transfer from the plasma to the component [12–14].

Despite the above-mentioned noticeable advantages, there are also some problems for plasma nitriding method. Some of them are as follows: (1) Need for handling of the process by skilled person, (2) The difficulty of processing big workloads with reproducible results, (3) Nonuniformity of the temperature distribution within the workload, (4) The overheating of small parts and (5) Surface damage by arcing [15].

References

1. Czerwinski F (ed) (2012) Heat Treatment-Conventional and Novel Applications. InTech, Rijeka
2. Prabhudev K H (2005) Handbook of Heat Treatment of Steels. McGraw-Hill, New Delhi
3. Totten G E (ed) (2007) Steel Heat Treatment Handbook, Metallurgy and Technologies. Taylor & Francis Group, Boca Raton
4. Krauss G (1990) Steels Heat Treatment and Processing Principles. ASM International, Ohio
5. Davis J R (ed) (2002) Surface Hardening of Steels. ASM International, Ohio
6. Chandler H (ed) (1995) Heat Treater's Guide: Practices and Procedures for Irons and Steels. ASM International, Ohio
7. Chattopadhyay R (2004) Advanced Thermally Assisted Surface Engineering Processes. Kluwer Academic Publishers, Dordrecht
8. (1991) ASM Handbook Vol. 4, Heat Treating. ASM International, United States of America
9. Negm N Z (2006) A study on rf plasma nitriding at a constant power in different H_2–N_2 mixtures at different temperatures. Mater Sci Eng B 129:207–210
10. Bouanis F Z, Jama C, Traisnel M, Bentiss F (2010) Study of corrosion resistance properties of nitrided carbon steel using radiofrequency N_2/H_2 cold plasma process. Corros Sci 52: 3180–3190
11. Totten G E (ed) (2007) Steel Heat Treatment Hand book, Equipment and Process Design. Taylor & Francis Group, Boca Raton
12. Rembges W, Oppel W (1993) Process control of plasma nitriding and plasma nitrocarburizing in industry. Surf Coat Technol 59:129-134
13. Xia Y, Wang S, Zhou F, Wang H, Lin Y, X Tao (2006) Tribological properties of plasma nitrided stainless steel against SAE52100 steel under ionic liquid lubrication condition. Tribol Int 39:635–640
14. Camps E, Becerril F, Muhl S, Alvarez-Fregoso O, Villagr′an M (2000) Microwave plasma characteristics in steel nitriding process. Thin Solid Films 373:293-298
15. Sun Y, Haruman E (2006) Effect of carbon addition on low-temperature plasma nitriding characteristics of austenitic stainless steel Vac 81:114–119

Chapter 2
Conventional DC Plasma Nitriding

As known before, the usual source for generating the plasma in plasma nitriding system is DC current (conventional or pulsed). This chapter just includes the results obtained by conventional DC plasma nitriding system on different grades of steels including stainless steels (austenitic, martensitic or precipitation hardening types), tool steels, and structural and constructional steels. This chapter consists of microstructural, mechanical, and corrosion properties obtained after conventional DC plasma nitriding of steels.

2.1 Microstructure

2.1.1 Austenitic Stainless Steels

The main phases present in the microstructure of plasma nitrided austenitic stainless steels are γ'-Fe_4N, ε-$Fe_{2-3}N$, and γ_N (expanded austenite). The expanded austenite (that is also called γ_N or s-phase) is formed by supersaturation of austenite FCC lattice by nitrogen atoms. The formation of this phase has two noticeable effects on XRD peaks of the nitrided layer: the broadening and shifting of the peaks to lower angles.

Here are the results of some experiments done to characterize the microstructure of plasma nitrided austenitic stainless steel.

Xu et al. [1] plasma nitrided AISI 304 austenitic stainless steel samples at temperatures of 400–420 and 500–520 °C, in atmosphere of 80 % N_2 + 20 % H_2, under pressure of 0.3 Pa and for 1 h. The microstructure of untreated sample (Fig. 2.1a) is composed of single phase γ-Fe. It can be seen from Fig. 2.1b that for 400 °C, all peaks are shifted to lower angles due to the formation of expanded austenite (γ_N) phase. It can be concluded from Fig. 2.1c that the layer produced at 500 °C is composed mainly of the α-ferrite and CrN phases.

H. Aghajani and S. Behrangi, *Plasma Nitriding of Steels*, Topics in Mining, Metallurgy and Materials Engineering, DOI 10.1007/978-3-319-43068-3_2

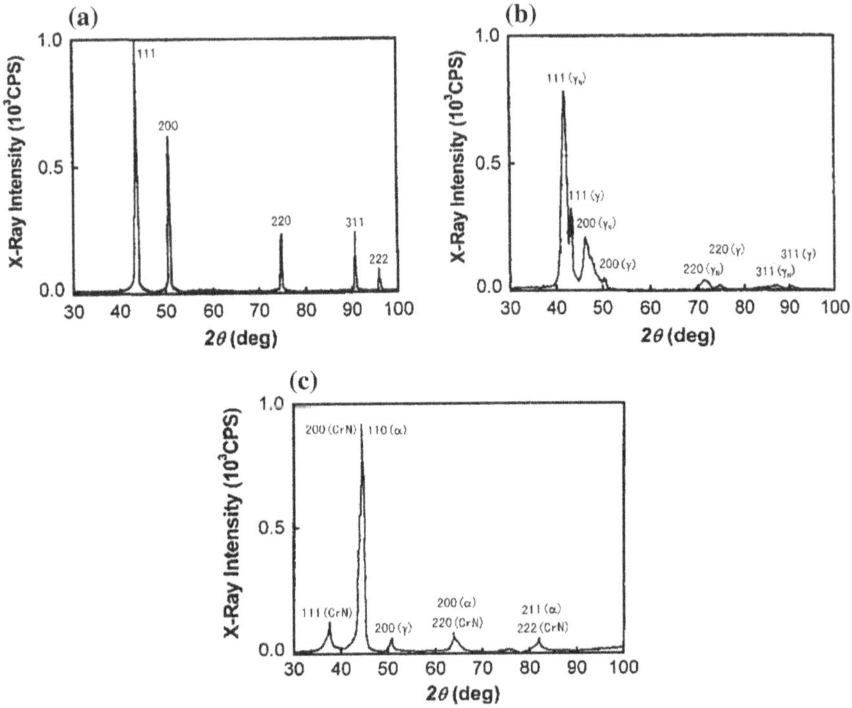

Fig. 2.1 XRD patterns of austenitic stainless steels. **a** Untreated, **b** after treatment at 400 °C and **c** after treatment at 500 °C [1]

According to TEM observations (Fig. 2.2), heavy stacking fault and high dislocation density regions have been observed for sample nitrided at 400 °C. Electron diffraction patterns indicated that an FCC phase had been produced. The lattice parameter of this phase was 0.372 nm, which was larger than that of the austenite (γ) parent phase. This phase is the mentioned expanded austenite (γ_N or s-phase). The formation of this phase induces stacking faults.

Figure 2.2 shows that, producing a bundle substructure is a result of nitrogen supersaturation and consequently decreasing of stacking fault energy.

A twinned relationship exists between bundles. These probably are deformation twins resulted from stress and strain which were produced from nitrogen supersaturation.

The microstructure of the plasma nitrided layer produced at 500 °C was made up of colonies displaying a lamellar structure. Each colony was reported to consist of α (ferrite or martensite) and CrN. With temperature enhancement, the diffusivity of substitutional atoms was increased. The bonding between chromium and nitrogen was stronger than that of other substitutional atoms with nitrogen, so that the CrN could nucleate and grow. Another suggestion for formation of CrN was decomposition of γ_N into α and CrN with increasing the temperature.

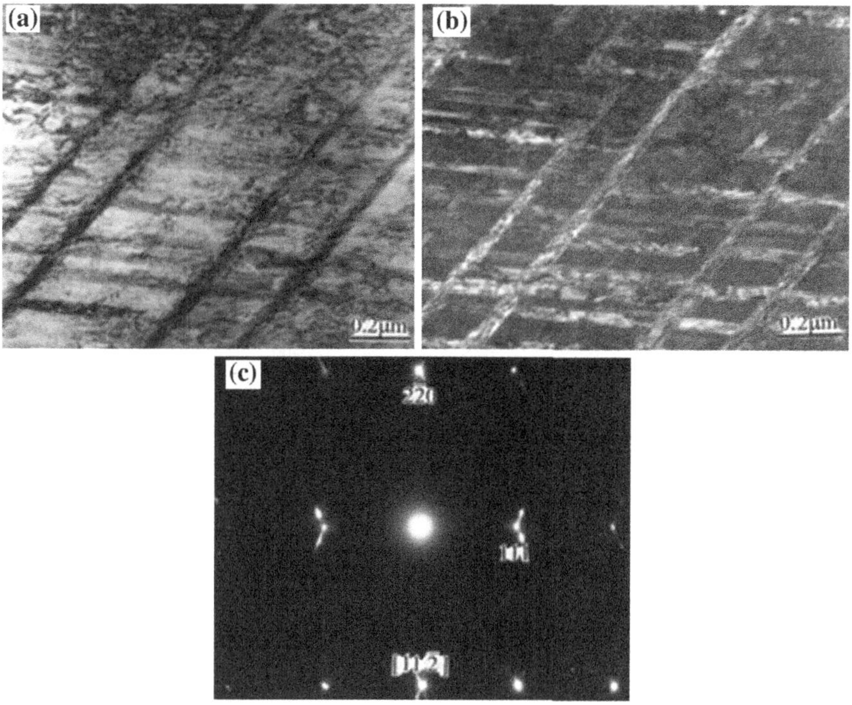

Fig. 2.2 TEM images showing stacking fault substructure of γ_N phase produced by plasma nitriding at 400 °C. **a** *Bright* field, **b** *dark* field and **c** electron diffraction pattern of (**a**) along [112] zone axis [1]

Wang et al. [2] plasma nitrided AISI 304L austenitic stainless steel with parameters listed in Table 2.1. There was no diffusion zone observed for none of the samples. However, it is obvious from Fig. 2.3 that the modified layer thickness is increased by temperature.

They also observed that, with an increase in temperature, the amount of γ_N decreased until 460 °C in which the γ_N completely disappeared. However, at 460 and 480 °C the CrN peaks started to get detected due to the precipitation of CrN which depletes the γ_N phase of chromium at 460 °C. This leads to the formation of a layer composed of expanded martensite (or austenite) and CrN.

At 480 °C, just CrN and austenite of substrate were detected because of above-mentioned decomposition process as well as the chemical reaction of expanded austenite (γ_N) with the incident nitrogen species during the plasma nitriding treatment.

Stinville et al. [3] plasma nitrided AISI 316L austenitic stainless steel samples at 400 °C, in atmosphere of 60 % N_2 + 40 % H_2, under pressure of 4.5 Pa and for 20 min–160 h. They detected γ_N phase. Also, increasing the temperature resulted in lowering the diffraction angle. They also reported an expansion of ~9 % in the direction normal to the surface and no expansion parallel to the surface for the

Table 2.1 The nitriding parameters used by Wang et al. [2]

Temperature (°C)	Duration (h)	Atmosphere $\left(\frac{N_2}{H_2}\right)$	Pressure (Pa)	Voltage (V)
350	4	$\frac{62}{240}$	600	520
400	4	$\frac{58}{196}$	600	520
420	4	$\frac{47}{201}$	600	520
460	4	$\frac{45}{180}$	600	520
480	4	$\frac{49}{199}$	600	520
420	4	$\frac{42}{172}$	400	520
400	4	$\frac{40}{161}$	400	520
350	4	$\frac{45}{186}$	400	520
400	4	$\frac{42}{168}$	200	520
420	8	$\frac{46}{191}$	600	520
400	8	$\frac{60}{246}$	600	520
350	8	$\frac{53}{220}$	600	520

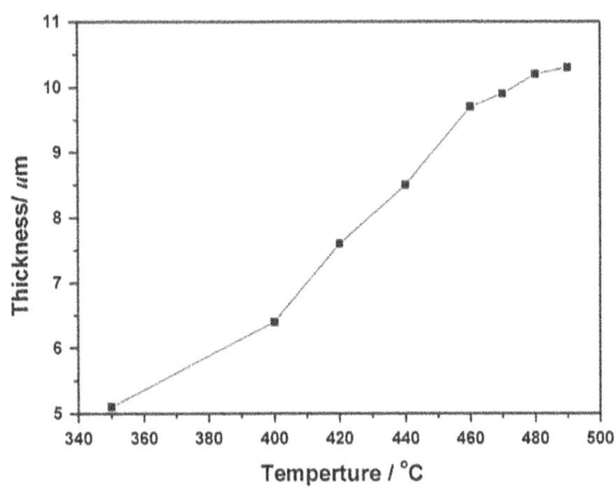

Fig. 2.3 Case depths of plasma nitrided AISI 304L samples treated at various temperatures [2]

nitrided layer. This led to the induction of high compressive stresses in the nitrided layer.

Li et al. [4] investigated plasma nitriding of AISI 316L austenitic stainless steel by applying these parameters: temperatures of 350, 400, 420, 450, and 480 °C, atmosphere of 20 % N_2 + 80 % H_2, pressure of 600 Pa and duration of 4 h. They observed no diffusion zone in any of the samples as reported by Wang et al. [2].

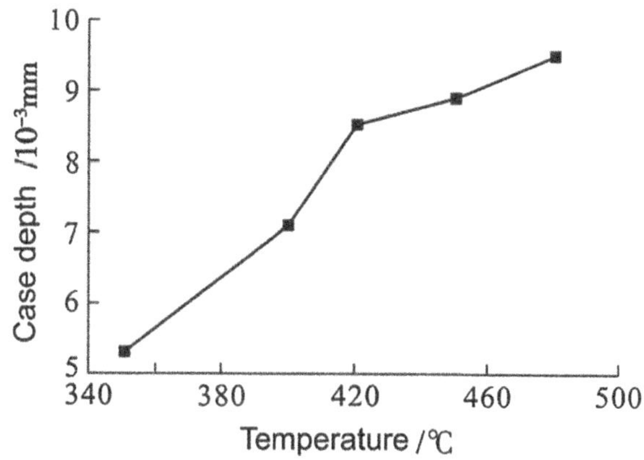

Fig. 2.4 Case depths of plasma nitrided AISI 316L austenitic stainless steel at various temperatures [4]

A dense case with thickness of approximately 5 μm produced at 350 °C which was less thicker than the layer formed at 480 °C. Besides, the layer formed at 480 °C showed a clear interface with substrate most probably because of the distinction between crystal structure of case and substrate. The thickness of treated layer is plotted in Fig. 2.4 as a function of temperature. It clearly can be seen in this figure that the thickness of the treated layer reaches a maximum value at 480 °C and the layer thickness shows an approximately linear behavior with temperature.

A high nitrogen concentration was seen at the surface of nitrided layer nitrided in various temperatures but at the same time. Due to a low diffusion rate at low temperatures, nitrogen penetration depth in the samples treated at 350 and 400 °C was less than that at temperatures higher than 420 °C.

In accordance with other researchers, they also saw that, the expanded austenite peaks were detected at lower temperatures and CrN was detected with increasing temperature. The reason for this action was given before. The broadening of γ_N peaks was probably due to the nitrogen gradient, residual stresses, and possible defect structure of the nitrided layers as mentioned before.

Egawa et al. [5] investigated the effect of some alloying elements on the microstructure of conventional DC plasma nitrided austenitic stainless steel samples nitrided with these parameters: temperatures of 400, 430, 440, and 470 °C, atmosphere of 80 % N_2 + 20 % H_2, pressure of 667 Pa and duration of 4 h. In this study different austenitic stainless steel samples have been used: AISI 304, AISI 316, JIS-SUS 304J3, AISI 321, and AISI 347.

They found that with increasing temperature, the formation of CrN phase increased (according to Fig. 2.5) and appeared as dark phases after etching with marble's reagent.[1] So, two critical temperatures were defined: T_1 as the critical temperature where the "dark" phases are formed along the grain boundary and T_2 as the critical temperature where the dark phases are formed at the intergranular

[1]mixture of equal parts of hydrochloric acid and $CuSO_4$ saturated aq. solution.

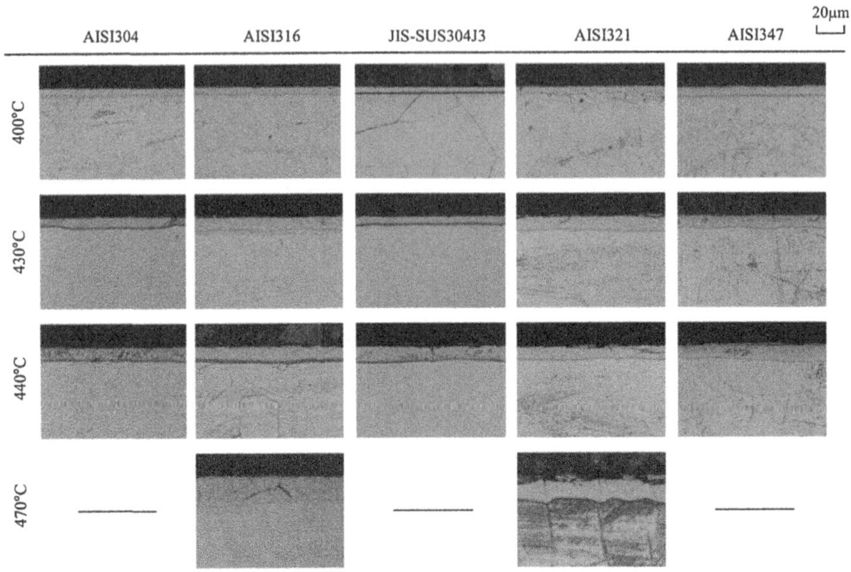

Fig. 2.5 Cross sections of the treated layers formed in various stainless steels after plasma nitriding at various temperatures for 4 h [5]

Table 2.2 Critical temperatures for formation of precipitations; at the grain boundaries (T_1) and at the intergranular regions (T_2) [5]

	304	316	304J3	321	347
T1 (°C)	430–440	430–440	430–440	430–440	430–440
T2 (°C)	430–440	460–470	430–440	460–470	430–440

regions. According to the authors, T_1 is independent of the chemical composition of substrate but T_2 is not constant (Table 2.2).

Molybdenum in AISI 316 and titanium in AISI 321 stabilize the structure of the expanded austenite. This stabilization is done by preventing the precipitation of chromium nitrides with attracting the surrounding nitrogen atoms. However, the formation of the nitride precipitation was not affected in JIS-SUS 304J3 because copper (existing in this steel) does not tend to attract nitrogen atoms. Despite the tendency of niobium in AISI 347 to attract nitrogen atoms, the T_2 temperature of AISI 347 was equal to that of AISI 304 because of combining with carbon.

They have noted that the behavior of layer thickness with temperature can be divided into two types: the first is when the layer grows slowly at low temperatures and then more rapidly above a critical temperature (AISI 304, J 304 and AISI 347) and the second is when an approximately linear increase in the layer thickness is shown with an increase in the temperature (AISI 316 and AISI 321). So in the first type, the precipitation of CrN accelerates the growth of the nitrided layer.

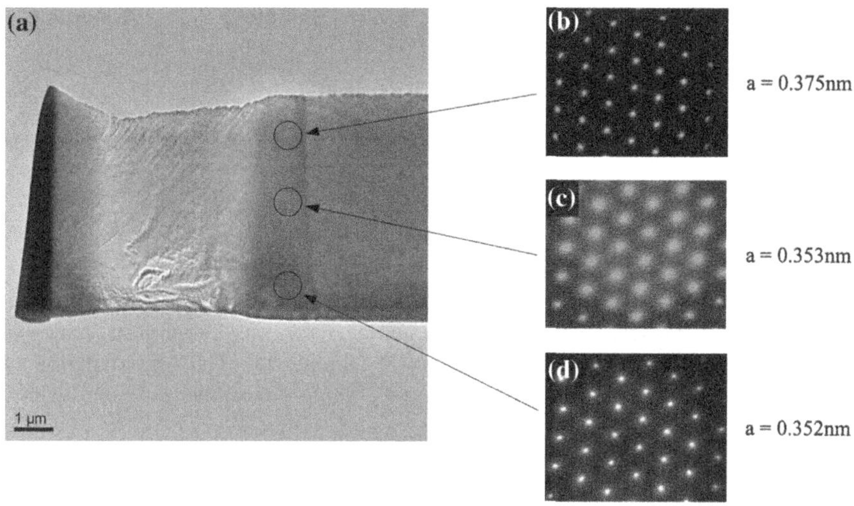

Fig. 2.6 **a** TEM image and **b–d** the selected area diffraction (SAD) patterns for the layers of the AISI 304 sample nitrided at 400 °C [5]

Figure 2.6 shows the increment of lattice parameters of the nitrided layers compared to the substrate austenite. Also, lattice parameter in the top part of the nitrided layer (Fig. 2.6b) was larger than that in the bottom region (Fig. 2.6d).

The formation of expanded austenite at low temperatures and CrN at high temperatures also was observed by Oliveira et al. [6]. They treated AISI 316L samples with applying these parameters: temperature of 350, 400, and 450 °C, atmosphere of 20 % N_2 + 80 % H_2, pressure of 667 Pa, and duration of 5 h. They observed the austenitic substrate and the growth of the nitrided layer, with the increasing nitriding temperatures. In the sample nitrided at 400 °C, the formation of only a white layer was observed. In the sample nitrided at 450 °C, emergence of the CrN precipitate in the white layer as well as a substantial increase of CrN precipitation in the sample nitrided at 500 °C, led to the formation of a second dark layer over the original white layer.

It was observed that the white layer was constituted mainly by the expanded austenite (γ_N). It was reported that in the case of sample plasma nitrided at low temperature (400 °C), γ_N peaks in XRD pattern appeared at lower angles compared to the substrate FCC austenite, indicating the larger lattice parameter of expanded austenite compared to the substrate austenite. This expansion was a result of the nitrogen supersaturation in austenite as mentioned by other authors before. Besides, CrN precipitates began to be detected with increasing temperature.

Olzon-Dionysio et al. [7] investigated the so-called "edge effect" phenomenon in plasma nitrided AISI 316L samples (Fig. 2.7). The thickness of all samples was 3.9 mm and diameter was 19.7 mm for samples A, B, and C. For sample D, the internal and external diameters were 19.7 and 31.6 mm, respectively. The plasma nitriding parameters were: temperature of 400 °C, atmosphere of 80 % N_2 + 20 % H_2, and duration of 4 h.

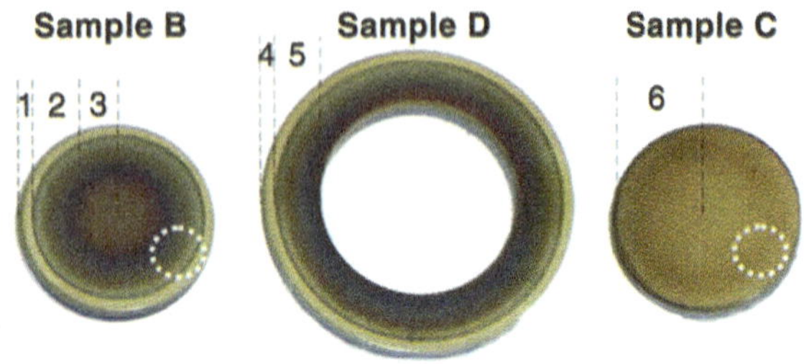

Fig. 2.7 Different regions of plasma nitrided samples according to their different colors [7]

According to Fig. 2.7, the borders of samples B and D (rings) were light gray, labeled 1 and 4, respectively. Toward the center, there was a region in shades of blue and brown, labeled 2 (sample B) and 5 (sample D). The central region of sample B is dark yellow, labeled as number 3. Sample C shows the same yellow shade and was labeled 6.

The depth of nitrided layer for sample C was 2.3 ± 0.2 μm. For sample B it was 1.5 ± 0.1 μm.

After plasma nitriding some phases such as γ_N, ε-$Fe_{2+x}N$ (hexagonal), γ' (FCC), and chromium nitrides were detected as well as γ phase. The peaks for γ_N were in lower angles compared to γ peaks and had broader shapes as a result of nitrogen gradient.

The amount of γ phase in sample B was more than sample C, suggesting a thinner nitrided layer for sample B. The γ_N phase of the sample with ring (B) was closer to the γ phase than the sample without ring (C). It indicates that the γ_N phase presents more nitrogen atoms for C than B.

The existence of Cr_2N phase was reported on the surface of sample B. The formation of this phase can explain why the γ_N phase of this sample presents less nitrogen atoms than sample C which is due to the consumption of part of the nitrogen to form this Cr_2N phase.

Although the nitriding temperature was 400 °C, it reached 420 °C in borders of sample because of edge effects. When reaching this temperature, the decomposition of metastable γ_N phase occurs due to reaching the solubility limit of nitrogen in the austenite structure and consequently chromium nitrides precipitates, which is relatively a stable phase. Consequently the γ_N phase decreased where samples B and D, which contain rings, showed smaller amounts of γ_N phase than C, without rings.

Nakajima et al. [8] studied the effects of solution treatment after plasma nitriding on the properties of the nitrided layer formed on AISI 304 austenitic stainless steel. The samples were solution treated at 1080 °C for 30 min followed by plasma nitriding at 500 °C for 8 h. The post-solution treatment was done at 1200 °C for 1 h.

No grain growth was reported by the authors because the grain size remained almost constant as follows: 65, 73, and 69 μm for untreated, nitrided, and nitrided-annealed samples, respectively. The nitrided layer was reported to be 50 μm thick and mainly constituted by CrN.

Wang et al. [9] studied the effects of treatment duration on properties of plasma nitrided AISI 304 austenitic stainless steel. The plasma nitriding treatment was done under pressure of 100 Pa and atmosphere of ammonia at about 420 °C for different durations.

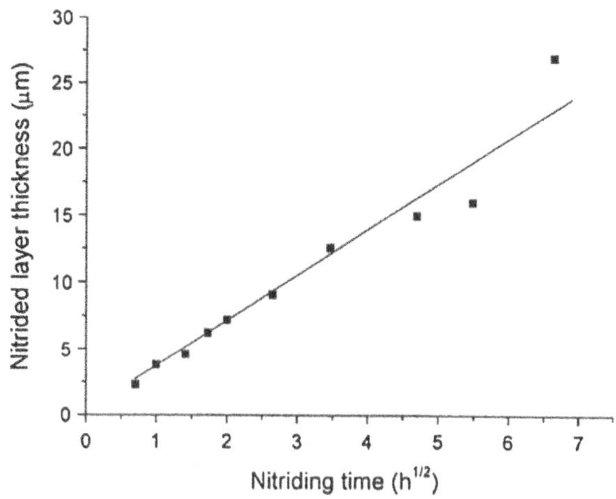

Fig. 2.8 The variations of nitrided layer thickness with nitriding time [9]

At durations lower than 8 h, the layer was featureless. However for longer treatment times of above 12 h, some defects or precipitations appeared in the nitrided layers. A linear relationship was observed between the layer thickness and second root of the duration. This can be seen in Fig. 2.8. For treatment durations of 30 min and 44 h, the thicknesses of the nitride layer were 3 and 27 μm.

The microstructure of the nitride layer was reported to be composed of expanded austenite and substrate austenitic phase. The amount of both phases increased by treatment duration.

Sun et al. [10] evaluated the effect of CH_4 addition to the nitriding atmosphere on the properties of nitrided AISI 321 austenitic stainless steel. Plasma nitriding was done at range of 400–430 °C for 20 h under atmosphere of 80 % N_2 + 20 % H_2. The addition of CH_4 was done with replacing H_2 with CH_4 with amount of 2–10 %. Three kinds of treatments were done: (1) nitriding in atmosphere of 80 % N_2 + 20 % H_2 for 5 h (NT), (2) treating in atmosphere of 95 % N_2 + 5 % CH_4 for 1 h and then nitriding in 80 % N_2 + 20 % H_2 atmosphere for 4 h (NT + NTC) and (3) treating in 95 % N_2 +5 % CH_4 atmosphere for 5 h (NTC).

A double layer structure was observed for NTC and NT + NTC treatments in which an N-enriched layer is formed on the top and a C-enriched layer is formed under it. The thickest C-enriched layer was produced by NTC treatment as shown in Table 2.3. It can also be seen in this table that the thickness of the N-enriched layer produced by NTC and NTC+NT treatments was almost the same, which was even larger than that produced by just NT. It is obvious that the addition of carbon had a

Table 2.3 Layer thickness and surface hardness obtained from various treatments [10]

Treatment	Total thickness (μm)	N-enriched layer thickness (μm)	Surface hardness ($HV_{0.1}$)
NT	7.8	6.2	890
NT + NTC	12	9.2	1340
NTC	14	9.2	1530

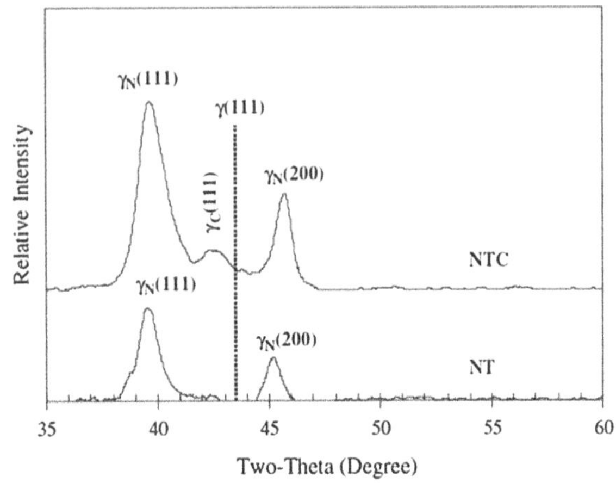

Fig. 2.9 XRD patterns from surface of the samples nitrided without carbon addition (NT treatment) and with carbon (NTC treatment) addition at 420 °C for 5 h [10]

positive effect on increasing the thicknesses of N-enriched and the C-enriched layers. The improvement in hardness is also observed in this table.

As seen before, the formation of expanded austenite is common phenomenon in plasma nitriding of steels at low temperatures. This also happened for NTC treatment in which γ_C has been formed as well as γ_N. This γ_C phase is formed by supersaturation of austenite with carbon leading to lattice expansion. This is clearly seen in Fig. 2.9. The measured lattice parameter for γ_N and γ_C phases was about 0.395 and 0.371 nm, respectively, showing a significant expansion of 10 % for N-enriched and 3 % for C-enriched layers compared to 0.359 nm for austenite phase.

Carbon addition to the nitriding atmosphere also has reported to improve the uniformity of the layer. The sample treated with only plasma nitriding showed a nonuniform layer thickness. However, carbon addition to the atmosphere significantly improved the uniformity of the layer.

The effects of CH_4 amount have also been investigated. With addition of 2 % CH_4, the modified zone was dominated by the N-enriched layer and had a thin C-enriched layer. Increasing the CH_4 level to 10 %, led to a further increase in thickness of the C-enriched layer and consequently the total layer thickness and the resultant N-enriched layer and C-enriched layer had similar thickness in this CH_4 amount. With a further increase in amount of CH_4 the C-enriched layer thickness increased, but with a reduction in the N-enriched layer thickness.

Borges et al. [11] studied the influences of different gas compositions of nitriding atmosphere as well as addition of C_2H_2 (acetylene) gas on properties of plasma nitrided AISI 304. Parameters of treatments were as follows: temperature of 400–600 °C, atmosphere of 95–100 % N_2, 1–5 % H_2, and 0–0.75 % C_2H_2, pressure of between 5 and 10 torr.

Figure 2.10 presents the results of XPS analysis which shows the concentration of chromium on the nitride surface. It can be seen that the concentration of chromium decreased with both hydrogen and acetylene concentration.

Fig. 2.10 XPS results showing the chromium concentration on the nitride surface for different acetylene concentrations [11]

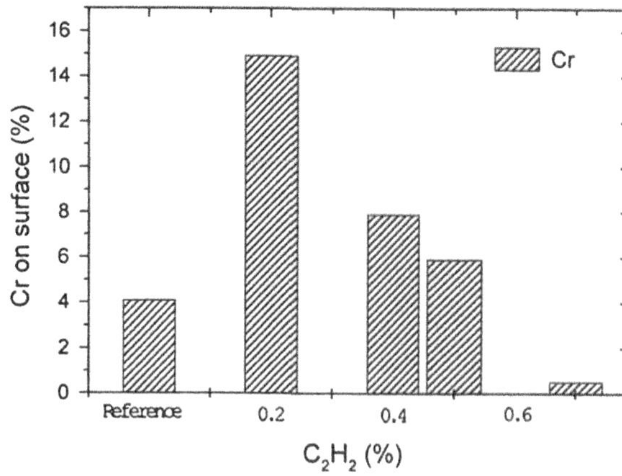

The amount of CrN formation was observed to directly be changed with the percent of hydrogen and acetylene in the nitriding atmosphere. However, the decrement of CrN was observed with temperature.

2.1.2 *Martensitic Stainless Steels*

Alphonsa et al. [12] studied plasma nitriding of AISI 420 martensitic stainless steel with these parameters: temperature of 530 °C, atmosphere of 20 % N_2 + 80 % H_2, pressure of 300–400 Pa, and duration of 20 h.

The only phase detected in untreated surface was α-Fe whereas for plasma nitrided sample, ε-Fe_3N was detected in addition to α-Fe. The presence of CrN and traces of Fe_4N, in addition to the Fe_3N were also reported. The higher carbon content (>0.15 %) in AISI 420 was probably responsible for predominancy of Fe_3N.

The case depth was 61 μm. The microstructure over the case depth was different from that within the first 2–5 μm from the surface which shows the predominancy of the ε-Fe_3N phase.

The XPS spectrum of Fe2p shown in Fig. 2.11 consists of single peak Fe2p3/2 and Fe2p1/2 at 707 and 720.17 eV. It seems from this figure that the processing has not resulted in surface oxidation because the values of binding energies and also the shape of spectra which is asymmetric shape are in accordance with metallic Fe. Also, the Fe oxidation results in some changes such as large chemical shift and satellite structure in Fe2p spectrum, which are not observed here.

N1s spectrum (Fig. 2.12) with a binding energy of 397 eV is a peak typical of a metal nitride and confirms the formation of nitride. This peak belongs to both CrN and Fe_3N.

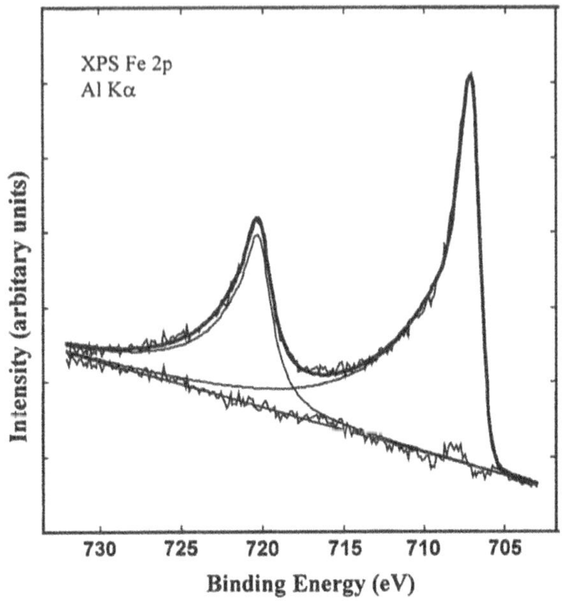

Fig. 2.11 Fe2p core-level XPS spectrum of plasma nitrided AISI 420 (*thin line*) along with the asymmetric Voigt function curve-fit (*thick line*). This figure clearly indicates that no oxidation occurred on the surface [12]

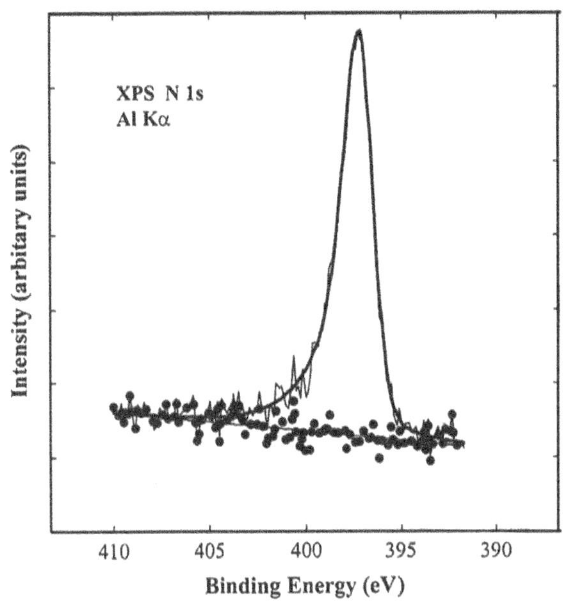

Fig. 2.12 N1s core-level XPS spectrum of plasma nitrided AISI 420 (*thin line*) superimposed with a single peak Voigt profile (*thick line*) indicating nitride formation [12]

Li et al. [13] investigated plasma nitriding properties of AISI 410 martensitic stainless steel. They did the treatment at temperatures of 420, 460, and 500 °C, in the atmosphere of 25 % N_2 + 75 % H_2, under pressure of 500 Pa for 20 h.

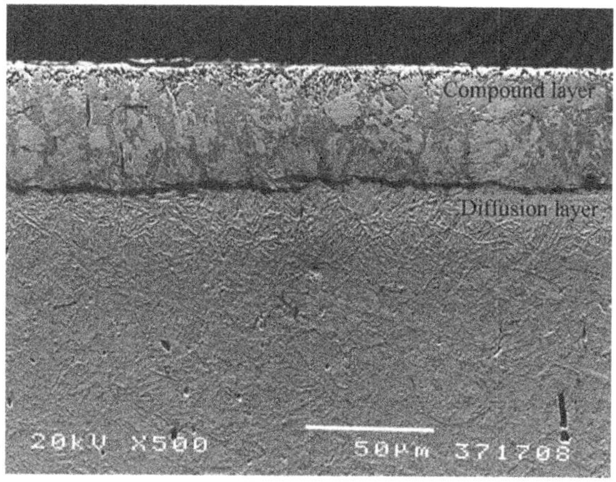

Fig. 2.13 Microstructure of AISI 410 martensitic stainless steel plasma nitrided at 420 °C [13]

The compound layer had a multiphase structure as indicated from composition variations in the layer in Fig. 2.13. The average thickness of the compound layer was 47, 86, and 135 μm for treating temperatures of 420, 460, and 500 °C, respectively. As seen in this figure, the microstructure in the diffusion zone is coarser than the substrate.

They reported that the untreated sample and the 420 °C nitrided surface consisted of mainly α-Fe and ε-$Fe_{2-3}N$, respectively. Small amounts of γ′-Fe_4N were also detected. The other phase detected was α_N or expanded martensite. This phase was formed by supersaturation of martensite by nitrogen atoms. No CrN peaks were detected by XRD analysis, indicating the prevention of the CrN formation in the surface of the AISI 410 steel when nitriding at 420 °C.

Increasing the treatment temperature to 460 °C, resulted in dominating of surface by γ′-Fe_4N, and reducing the amount of ε-$Fe_{2-3}N$. CrN precipitates were also detected in the nitrided surface. Because of overlapping the peaks of α_N with those of CrN or ε, α_N could not be positively determined.

At 500 °C, the amount of γ′-Fe_4N in the nitrided surface was further increased, but the amount of ε-$Fe_{2-3}N$ became minor. It was believed that at 500 °C nitrogen in the α_N phase dissoluted and combined with Cr in the steel to form stoichiometric CrN precipitates. This reduced the amount of nitrogen in solid solution, diminished the lattice expansion, and thus, shifted the XRD peak back to the position for α-Fe.

Pinedo et al. [14] studied plasma nitriding of AISI 420 martensitic stainless steel. They did the treatment with these parameters: temperatures of 480, 500, 520, 540, and 560 °C, atmosphere of 75 % N_2 + 25 % H_2, pressure of 250 Pa, and duration of 4 h.

They observed that increasing temperature resulted in thickening of the compound layer. No grain boundaries in nitrides network in the diffusion zone were observed. Also a clear interface between the diffusion zone and the substrate was observed for all nitriding temperatures.

The compound layer was found to consist of γ'-Fe_4N, ε-$Fe_{2-3}N$, and CrN nitrides.

Wu et al. [15] applied new and modified plasma nitriding methods to improve the properties of AISI 420 martensitic stainless steel. Cyclic plasma oxynitriding and cyclic plasma nitriding catalyzed by rare earth La were performed and compared with conventional plasma nitriding.

Conventional plasma nitriding was done in atmosphere of 70 % N_2 + 30 % H_2, at temperature of 560 °C for 25 h.

For cyclic plasma oxynitriding, some air entered in treatment atmosphere with amount of less than 5 % but the ratio of N_2/H_2 remained 7:3. For cyclic plasma nitriding catalyzed by La, four blocks of La were put directly in the four corners of the plasma nitriding furnace. The treatment cycle was as follows: nitriding at 560 °C for 3 h, cooling to room temperature in 30 min in an atmosphere of pure H_2, repeating of this stage for three times.

It was observed that the modified layer of the samples treated by conventional nitriding and cyclic plasma nitriding catalyzed by La, consisted of two layers of compound layer and diffusion zone in contrast with cyclic plasma oxynitrided samples that just had diffusion layer. The nitrided depths of the samples treated using conventional plasma nitriding, cyclic plasma oxynitriding, and cyclic plasma nitriding catalyzed by rare earth La in four cycles were 0.25, 0.35, and 0.64 mm, respectively.

The advantage of cyclic plasma oxynitriding technique was formation of a thicker layer in a shorter time so the role of some air was evaluated to be positive for accelerating the nitriding process. With developing the reaction between iron and oxygen in the plasma state, the formation of a loose and porous oxides on the surface was accelerated leading to absorbing a large amount of nitrogen by these active sites during the deoxidization process. The mentioned process led to a high nitrogen concentration in the nitrided case. Therefore a thicker nitrided case has been obtained with speeding up the nitrogen transportation.

Using La as a catalyst also increased the nitrogen diffusion on the surface. This occurred as a result of ionizing the La atoms and their sedimentation on the surface of the sample and activation of the surface by lattice distortion for increasing the diffusion of nitrogen atoms.

During cooling from nitriding temperature, the nitrogen-enriched ε phase has been transformed into less nitrogen containing γ' phase. This action has increased the interfaces and consequently the channels for diffusion of nitrogen. With reheating the sample, more nitrogen atoms diffused into the sample and the nitride layer grew thicker.

After treatment, all of the nitride layers were consisted of γ', CrN and α. Some ε phase was also detected for the samples nitrided by conventional plasma nitriding in 25 h and cyclic plasma nitriding catalyzed by rare earth La in four cycles. For the sample treated by the cyclic plasma oxynitriding, traces of Fe_3O_4 have been also detected.

2.1.3 Precipitation Hardening Stainless Steels

Esfandiari et al. [16] studied plasma nitriding behavior of A286 (AISI 600) precipitation hardening stainless steel with following parameters: temperatures of 350, 370, 390, 420, 460, and 500 °C, atmosphere of 25 % N_2 + 75 % H_2, pressure of 500 Pa, and durations of 5, 10, 20, and 30 h.

They reported that the layer thickness varied from 1.2 μm (350 °C/5 h) to 60 μm (500 °C/30 h). Treating at 350 °C for 10 h resulted in transformation of the austenite phase in the modified layer into expanded austenite. At 420 °C in short time (5 h), surface of the nitrided samples mainly consisted of s-phase. Further increase in time (10 h and 20 h) caused γ'-Fe_4N and Cr_2N to start to form. Cr_2N phase was also formed at 500 °C. Therefore, it can be concluded that in precipitation hardening stainless steels, chromium nitride starts to form with increasing temperature as observed in case of austenitic stainless steels.

At a constant temperature (specially low treating temperatures), a linear relationship was found between the square root of time and the thickness of the layer. Figure 2.14a shows the variations of the nitride layer thickness with the treatment time. The linear growth of the nitrided layer with the square root of time confirms the governing of layer growth by nitrogen solid diffusion.

Also as can be seen in Fig. 2.14b, the growth of nitrided layers can be divided into two regions: low-temperature (350–390 °C) region with slow growth and high-temperature (420–500 °C) region with a fast growth. The activation energies for these two regions are 115 and 104 kJ/mol, respectively.

Sobiecki et al. [17] did an investigation on plasma nitriding of H9S2 stainless steel (with chemical composition shown in Table 2.4) with conditions listed in Table 2.5.

In accordance with other researchers, they also found the dependence of the diffusion nitrided layer thickness on both the temperature and the duration of the process. The thickness of the layers produced by processes 1 and 3 was the same. According to their results, the composition of nitriding atmosphere did not change the thickness of the nitrided layer.

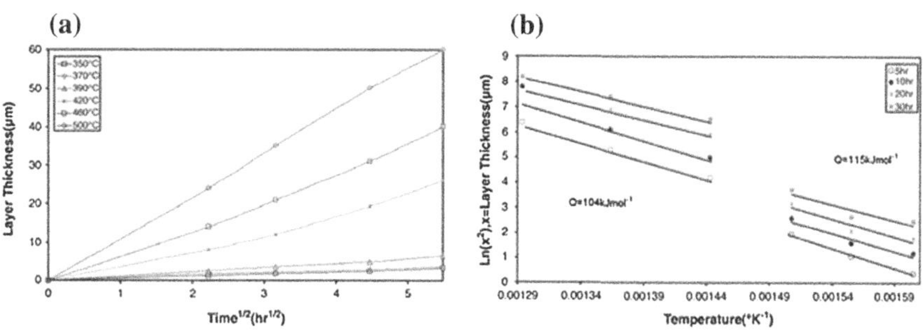

Fig. 2.14 Dependence of nitriding kinetics of A286 on **a** time and **b** temperature [16]

Table 2.4 Chemical composition of H9S2 steel [17]

C (%)	Mn (%)	Si (%)	Cr (%)	Ni (%)	Fe (%)
0.35–0.45	0.7	2	8–10	0.6	Balance

Table 2.5 Plasma nitriding parameters used by Sobiecki et al. [17]

No.	Temperature (°C)	Duration (h)	Atmosphere $\left(\frac{N_2}{H_2}\right)$
1	450	6	$\frac{1}{2}$
2	520	6	$\frac{1}{2}$
3	570	3	$\frac{1}{2}$
4	620	3	$\frac{1}{2}$
5	650	3	$\frac{1}{2}$
6	550	3	$\frac{1}{1}$
7	550	3	$\frac{2}{1}$
8	550	3	$\frac{1}{4}$

Li et al. [18] studied plasma nitriding of 17–4 precipitation hardening stainless steel. They applied these parameters: temperatures of 350, 400, 420, 450, and 480 °C, atmosphere of 20 % N_2 + 80 % H_2, pressure of 600 Pa, and duration of 4 h.

As can be seen in Fig. 2.15a, a 5 μm bright layer formed on the sample which was plasma nitrided at 350 °C for 4 h. However, for both of the samples treated with 420 °C/4 h and 480 °C/4 h conditions, an approximately 10 μm thick dark layer was observed (Fig. 2.15b, c). This indicated the susceptibility of this type of layer (from temperature above 400 °C processes) to etching in nital.

Nitrogen concentration profile across nitrided layers is shown in Fig. 2.16. At high temperatures, defect recombination (Frenkel pairs) resulted in a reduction in the surface stress and favoring nitrogen diffusion. Consequently, the highest nitrogen content of these nitrided layers was measured to be 12 wt% at 480 °C. Therefore, the nitrogen concentration at the surface of nitride layers formed with nitriding at 480 °C was higher and nitrogen penetration depth was deeper in comparison to the samples nitrided at temperatures less than 480 °C. Moreover, nitriding at increased temperature led to nitriding depths of more than 10 μm.

They have found that untreated sample mainly consisted of α phase. With the inward diffusion of the nitrogen, the FCC grain lattice was supersaturated by nitrogen to such extent that the transformation of alpha (α) into s-phase occurs. Therefore, at 350 and 400 °C, the main constituents of microstructure were a metastable "s-phase," substrate alpha (α) and the α_N (expanded martensite). However, as seen before about austenitic stainless steels, at temperatures above 420 °C the s-phase peaks disappear. This was probably due to the depletion of the

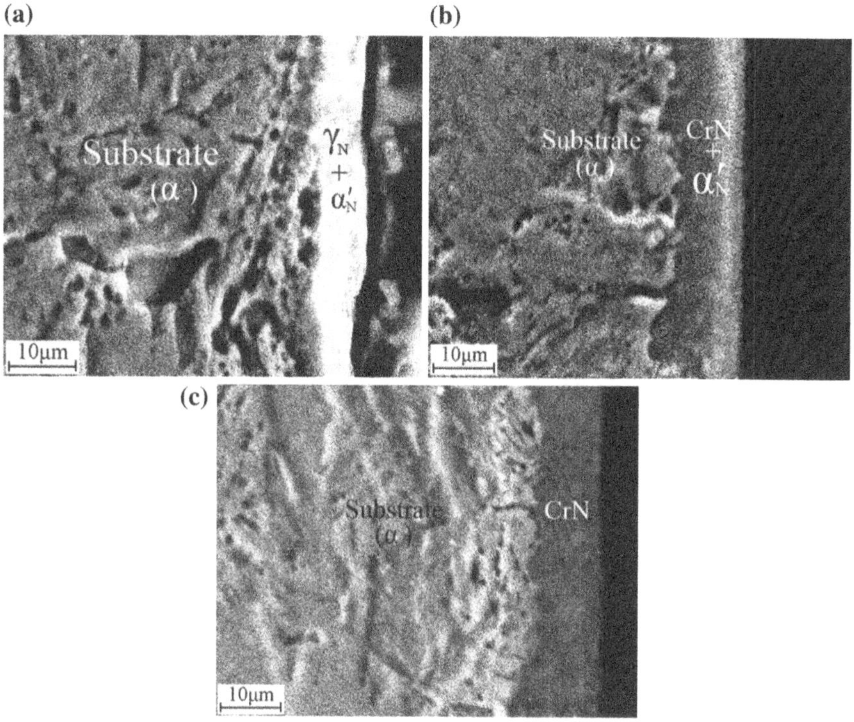

Fig. 2.15 Cross-sectional SEM micrographs of samples nitrided at various temperatures of **a** 350 °C, **b** 420 °C, and **c** 480 °C [18]

Fig. 2.16 Nitrogen concentration profile achieved at different temperatures for 4 h [18]

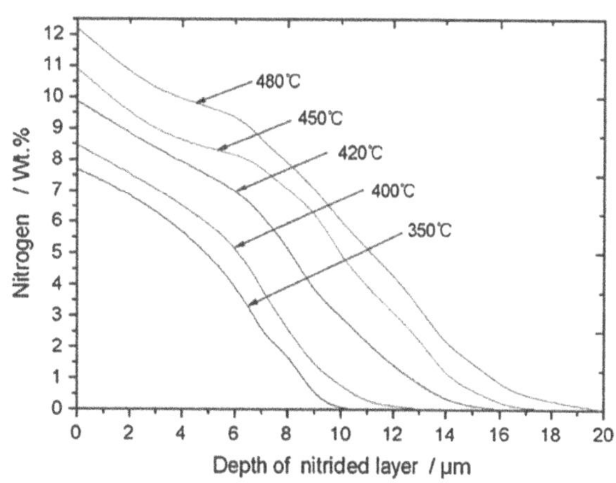

expanded phase of chromium at 420 °C resulted by CrN precipitation which results in the formation of a mixture layer of α_N and CrN (s-phase → α_N + CrN). Completely disappearing of α_N was possibly due to the decomposition of α_N into alpha (α) and CrN. Furthermore, at temperature ≥ 450 °C the alpha (α) peak sharpened and returned to the unnitrided diffraction angle probably because of structure relaxation produced by the precipitation ($\alpha_N \rightarrow \alpha + CrN$).

Dong et al. [19] studied plasma nitriding of 17-4PH martensitic stainless steel and found almost the same results found by Li et al. [18]. They applied the following conditions: temperatures of 350, 370, 390, 420, 460, and 500 °C, atmosphere of 25 % N_2 + 75 % H_2, pressure of 500 Pa, and durations of 10, 20 and 30 h.

For samples nitrided under 420 °C, a bright layer was observed. This was due not to have been attacked by etchant solution. However, the upper part of the layer treated at 420 °C for 20 h was slightly etched, indicating the nitrides precipitation during the treatment. With nitriding at 500 °C or above, the nitrided case became dark and grain boundaries in the treated case were preferentially etched. It should be noted that further increasing the nitriding temperature or time resulted in the development of the dark layer toward the layer/core interface.

The untreated sample mainly constituted by martensite and retained austenite (γ). High-temperature (≥ 460 °C) treated samples mainly consisted of chromium nitride (CrN) and γ'-Fe_4N. However, low-temperature nitrided samples were characterized by highly overlapped peaks and it was very difficult to characterize overlapped peaks of potential phases (γ'-Fe_4N, s-phase, α_N and CrN). Therefore, TEM was also used for further studies.

TEM observation of untreated 17-4PH stainless steel revealed the dominance of the lath martensite phase with the length of about 2–2.5 μm and width of 0.3–0.9 μm. A typical microstructure and corresponding SAD of $b = [100]_\alpha$ pattern are shown in Fig. 2.17. There were traces of retained austenite along the boundaries of martensite laths.

The maintained martensite features as well as micro twins and slip lines across martensite laths were observed for 350 °C/10 h treated sample. Also, the calculated lattice parameter ($a = 0.290$ nm) for this sample was larger than for the untreated samples ($a = 0.287$ nm). This difference can be ascribed to the supersaturation of nitrogen in the martensite. Also GDS (Glow Discharge Spectroscopy) results revealed the abundance of nitrogen amount in the nitrided layer compared to the solubility of nitrogen in the martensite lattice. Therefore, it can be concluded that the nitrided layer formed at temperatures below 420 °C is a nitrogen supersaturated martensite layer rather than a compound layer.

Microstructural observations revealed the existence of isolated s-phase grains in all the samples plasma nitrided at 420 °C and lower temperatures for 10 h. They also observed that α_N surrounded the s-phase and the size and distribution of the s-phase were similar to that of the retained austenite. From these findings and observations, it was concluded that the formation of s-phase in the martensitic 17-4PH stainless steel was resulted by the conversion of the retained austenite in the initial material. Therefore, the formation of s-phase was not as a continuous layer in plasma nitrided layers.

According to the results of TEM investigation (Fig. 2.18), when 17-4PH stainless steel was plasma nitrided at temperatures above 420 °C or at 420 °C for

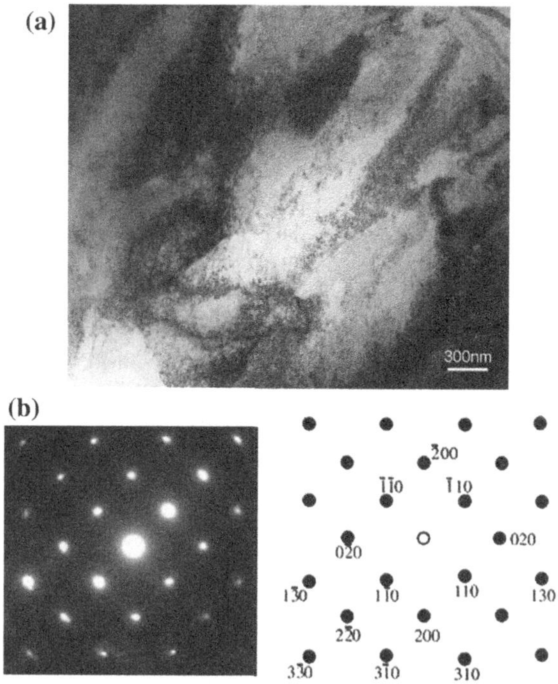

Fig. 2.17 **a** *Bright* field TEM micrograph and **b** corresponding SAD pattern, b = [100], of untreated 17-4PH stainless steel [19]

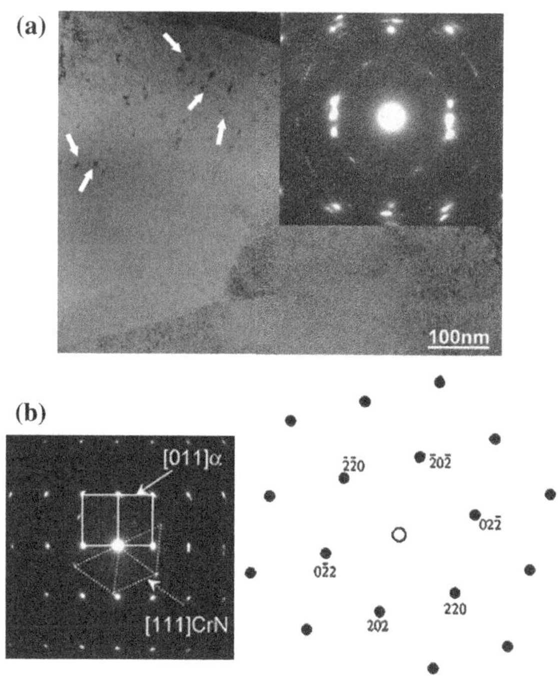

Fig. 2.18 **a** *Bright* field TEM micrograph and **b** corresponding SAD pattern from the CrN phase in 420 °C/20 h treated sample [19]

more than 10 h, the decomposition of s-phase into CrN and α-Fe occurred simultaneously with precipitation of CrN from α_N.

2.1.4 Tool Steels

Birol [20] investigated plasma nitriding behavior of the X32CrMoV33 hot work tool steel with these conditions: temperature of 520 °C, atmosphere of 100 % NH_3, pressure of 100 Pa, and duration of 12 h.

The compound layer had the thickness of 6 μm and consisted of ε-$Fe_{2-3}N$ phase.

Nouveau et al. [21] studied plasma nitriding of 90CrMoV8 tool steel at temperatures of 460, 500, and 520 °C, in atmosphere of 80 % N_2 + 20 % H_2, under pressure of 500 Pa and for 6, 8, 10, and 12 h.

The Fe and all the iron nitride peaks were broadened and shifted to lower angles with treatment duration (with respect to untreated sample).

With increasing the treatment duration, γ′ (110) and (200) peaks were broadened. The ε phase in all durations was present with high intensity as observed by Birol [20].

It was observed that in a fixed treatment duration, the Fe (110) peak widened and shifted toward the lower angles with respect to untreated steel ones, regardless of the process temperature. It can be implied that the temperature was less effective than duration. Also it was obvious that, the higher the temperature, the lower the angles and no trace of chromium nitrides was observed.

da Silva Rocha et al. [22] investigated plasma nitriding of AISI M2 high-speed steel. They applied these conditions: temperatures of 350, 400, 450, and 500 °C, atmosphere of 5 and 76 % N_2 + 95 and 24 % H_2, pressure of 500 Pa, and duration of 30 min.

The untreated sample was constituted by the α-iron as well as carbides of the tool steel. A gas composition of 76 vol.% N_2 in the plasma resulted in the formation of a compound layer at 400 and at 500 °C composed of the ε-phase. At 500 °C alloying elements nitrides were formed and the primary and secondary carbides of steel transform to carbonitrides or even nitrides.

For lower nitrogen content in the gas, higher intensity of α-Fe, no compound layer at 500 °C, no ε-nitride, and some γ′-nitride were observed.

Figure 2.19 confirms XRD results. As can be seen in Fig. 2.19, no continuous compound layer was formed at the surface of samples nitrided in atmosphere of 5 vol.% N_2. However, plasma nitriding in the atmosphere of 76 vol.% N_2 resulted in the formation of compound layer at all temperatures. As can be seen from Fig. 2.19 and Table 2.6, 5 vol.% N_2 led to the formation of a thin diffusion zone only. The depth of the diffusion zone increased from 4 to 6 μm to about 16–20 μm with temperature. However, with 76 vol.% N_2, both compound layer thickness (from 0.5 to 2.3 μm) and depth of diffusion zone increased with temperature (from 8 μm to about 30 μm).

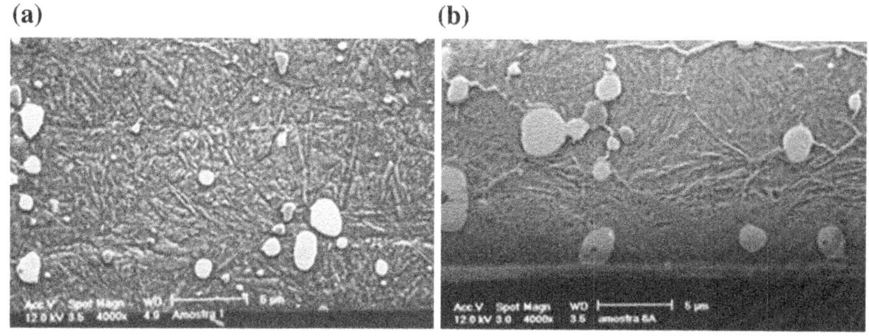

Fig. 2.19 SEM Cross-sectional images of a sample plasma nitrided at 500 °C for 30 min in atmosphere of **a** 5 vol.% N_2 and **b** 76 vol.% N_2 [22]

Table 2.6 Compound layer and diffusion zone thicknesses of samples plasma nitrided at different temperatures and with different gas compositions [22]

Parameters	Compound layer (μm)		Diffusion zone	
	Metallography	GDOS	Metallography	GDOS
350 °C, 5 % N_2	–	–	6.0	4.0
350 °C, 76 % N_2	0.5	0.4	8	4.0
400 °C, 5 % N_2	–	Not measured	6.5	Not measured
400 °C, 76 % N_2	0.7	Not measured	6.5	Not measured
450 °C, 5 % N_2	–	Not measured	12.5	Not measured
450 °C, 76 % N_2	1.3	Not measured	14	Not measured
500 °C, 5 % N_2	–	–	16.0	23.4
500 °C, 76 % N_2	2.3	2.8	24.0	32.6

Pessin et al. [23] studied plasma nitriding of AISI M2 high-speed steel with parameters presented here: temperatures of 450 and 500 °C, atmosphere of 25 % N_2 + 75 % H_2, pressures of 200, 400 and 900 Pa, and duration of 60 min.

Table 2.7 summarizes the results of metallographic analysis and X-ray diffraction. The X-ray diffraction patterns for two of the conditions used are shown in Fig. 2.20. Although the microstructural analysis of all samples did not reveal a compound layer, it can be seen from Table 2.7 and Fig. 2.20 that the γ'-phase was formed at 500 °C and the ε-phase was observed for both temperatures, which indicated the formation of compound layer. It can be observed that the nitride formation ratio ε/γ' decreases when temperature is increased.

Increasing of pressure from 200 to 400 and 900 Pa at 500 °C resulted in the formation of γ'-Fe_4N and ε-$Fe_{2-3}N$ phases, although only ε-nitrides were observed at 450 °C.

According to Table 2.7, it is obvious that samples nitrided at 400 Pa exhibited deeper case depths than those nitrided at 200 and 900 Pa. This can be due to better plasma distribution and nitriding efficiency in this pressure.

Table 2.7 Iron nitrides identified (X-ray analysis) and layer depth (GDOS test) [23]

	Temperature					
	450 °C			500 °C		
	200 Pa	400 Pa	900 Pa	200 Pa	400 Pa	900 Pa
Compound layer composition	–	ε	ε	–	ε and γ′	ε and γ′
Layers depth (μm)	22	36	31	31	57	32

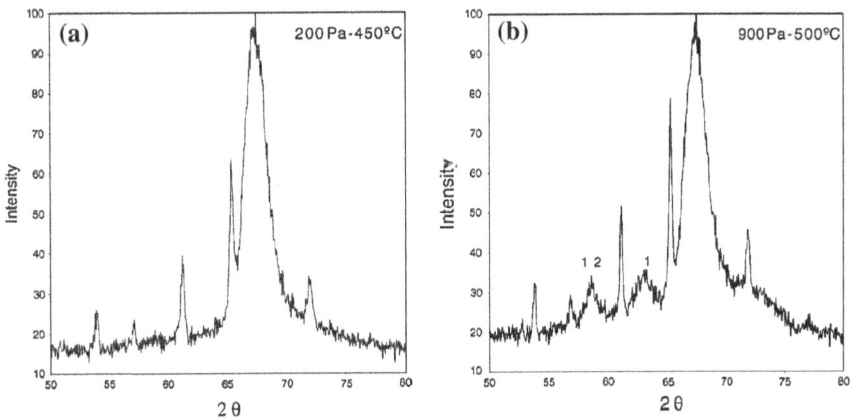

Fig. 2.20 XRD patterns showing **a** the absence of compound layer and **b** the characteristic peaks of ε (1) and γ′ (2) nitrides [23]

2.1.5 *Structural and Constructional Steels*

Li et al. [24] studied plasma nitriding of 709M40 low alloy steel with applying these parameters: temperature of 500 °C, atmospheres of 25 % N_2 + 75 % H_2 and 65 % N_2 + 35 % H_2, pressure of 500 Pa, and durations of 3, 10 and 40 h.

The samples were treated at 500 °C for 10 h with two nitrogen potentials. Both of the conditions produced a similar case. The compound layer was mainly consisted of γ′-Fe_4N with traces of ε-$Fe_{2-3}N$. The average thickness of the compound layer was 4.3 μm (for 25 % N_2) and 7.5 μm (for 65 % N_2).

Karaoğlu [25] investigated plasma nitriding of AISI 5140 low-alloy steel. The applied parameters are presented in Table 2.8.

The nitrided samples consisted of a compound layer and a diffusion zone.

As can be seen in Table 2.9, increasing time and/or temperature increased the thickness of the diffusion layer by increasing the nitrogen diffusion through the layer. It seems that, increasing the nitrogen content in gas mixture resulted in decreasing the case depth. Also it can be indicated from Table 2.9 that the process temperature is the most effective parameter on the thickness of compound layer.

It was reported that at the same temperature, the amount of ε-phase decreased but that of γ′ increased with time. Also increasing the process temperature caused

Table 2.8 Plasma nitriding conditions used by Karaoğlu [25]

Group	Temperature (°C)	Duration (h)	Atmosphere $\left(\frac{N_2}{H_2}\right)$	Pressure (Pa)	Voltage (V)
1.500.20	500	1	$\frac{20}{80}$	1000	400–600
4.500.20	500	4	$\frac{20}{80}$	1000	400–600
12.500.20	500	12	$\frac{20}{80}$	1000	400–600
4.450.20	450	4	$\frac{20}{80}$	1000	400–600
4.550.20	550	4	$\frac{20}{80}$	1000	400–600
4.500.60	500	4	$\frac{60}{40}$	1000	400–600

Table 2.9 Maximum case depth and compound layer thickness values obtained with applying different process parameters [25]

	Specimen group					
	1.500.20	4.500.20	12.500.20	4.450.20	4.550.20	4.500.60
Surface hardness (HV)	760	986	862	943	810	940
Case depth (μm)	166	320	496	247	424	306
Compound layer (μm)	2.5	5.5	7	2	11	7.5

an increase in the amount of γ' phase but reduced the amount of ε phase. Lowering the nitrogen content in the nitriding atmosphere led to an increase in the formation of γ' but a reduction in formation of ε.

Karakan et al. [26] studied plasma nitriding of AISI 5140 steel by means of different gas mixtures of 10 % H_2 + 90 % N_2, 10 % Ar + 90 % N_2, 90 % Ar + 10% N_2, 90 % H_2 + 10 % N_2, 10 % Ar + 10 % H_2 + 80 % N_2, 80 % Ar + 10 % H_2 + 10 % N_2, 10 % Ar + 80 % H_2 + 10 % N_2. The treatment temperature and duration were 450 °C and 4 h. Figures 2.29 and 2.30 show XRD results for AISI 5140 low-alloy steel plasma nitrided under various gas mixtures. As seen clearly in these figures, multiphase (γ and ε) iron nitrides formed on the surface of plasma nitrided low-alloy steel.

Figure 2.21 shows the XRD results of the plasma nitrided specimens at gas mixtures of 10 % H_2 + 90 % N_2 and 10 % Ar + 90 % N_2. For as much as the intensity of ε phase was not high, α-Fe phase was clearly observed when Ar–N_2 mixtures were used as nitriding atmosphere. Gas mixture of 10 % Ar + 90 % N_2 led to the formation of a thin compound layer because the amount of Ar in the gas mixture was not sufficient to create diffusion. It was observed that the presence of at least 20 % Ar in dual gas mixture is necessary to form a compound layer and to increase surface hardness. However, using 10 % H_2 gas mixture formed a compound layer because the ionization energy of H_2 is lower than that of Ar.

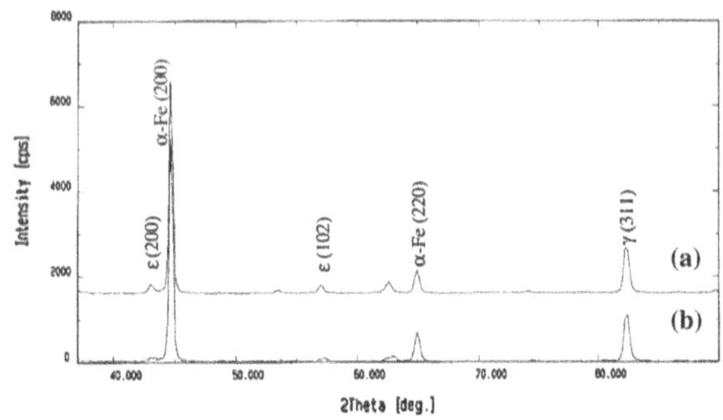

Fig. 2.21 XRD patterns of plasma nitrided AISI 5140 steel for gas mixtures of 10 % H_2 + 90 % N_2 (**a**) and 10 % Ar + 90% N_2 (**b**) [26]

Figure 2.22 illustrates the XRD results of the samples plasma nitrided at gas mixtures of 90 % H_2 + 10 % N_2 and 90 % Ar + 10 % N_2. While the intensity of ε decreased for Ar–N_2 mixtures, the intensity of γ increased for H_2–N_2 mixtures.

When the amount of Ar in the gas mixture increased, the intensity of compound layer phases increased.

Mahboubi et al. [22] investigated plasma nitriding of DIN 1.6959 low alloy steel. They have applied these conditions: temperatures of 350, 400, 450, 500, and 550 °C, atmosphere of 75 % N_2 + 25 % H_2, pressure of 400 Pa, and duration of 5 h.

According to Fig. 2.23, the compound layer is observable for all samples. Plasma nitriding at 550 °C led to a more uniform interface between compound layer and diffusion zone. However for samples nitrided at 350 and 450 °C, the compound layer extended into the grain boundaries at some areas along the interface. This can be due to the grain boundary and volume diffusion coefficients which are temperature dependent. At low treatment temperatures, the mobility of nitrogen atoms was low and penetration was done more easily at less compact diffusion routes such as grain boundaries. Therefore, nitrogen atoms diffusing along the boundaries will be able to penetrate much deeper than atoms that only diffuse through the more compact areas such as grains (lattice diffusion). Increasing treatment temperature

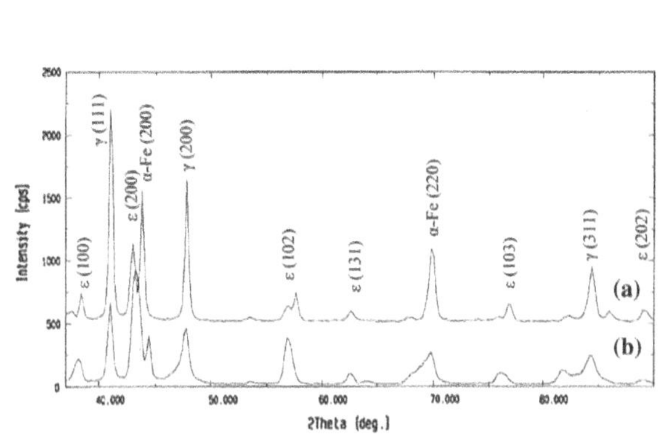

Fig. 2.22 XRD patterns of plasma nitrided AISI 5140 steel for gas mixtures of 90 % Ar + 10 % N_2 (**a**) and 90 % H_2 + 10 % N_2 (**b**) [26]

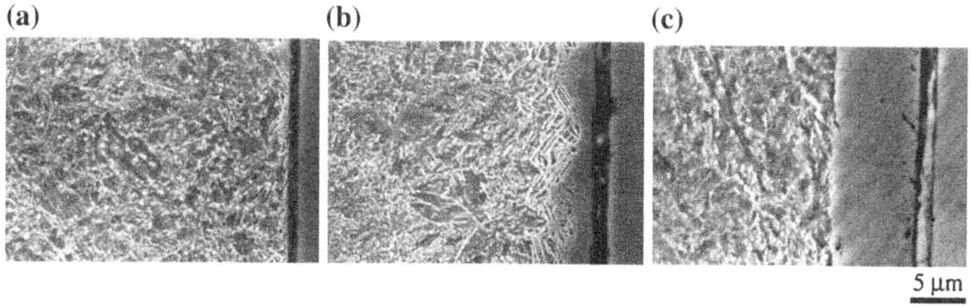

Fig. 2.23 SEM images of the surface layers of the samples plasma nitrided at various **a** 350, **b** 450, and **c** 550 °C [27]

led to the increment of the nitrogen atoms mobility. Thus, lattice diffusion dominated and the penetration of nitrogen atoms through grains occurred more easily, leading to the formation of a more uniform compound layer.

They have observed that the thickness of compound layer and diffusion zone generally increased with increasing temperature because the growth of the compound layer is controlled by diffusion of nitrogen through this layer.

The compound layer of all samples mainly consisted of ε and γ′ nitrides. Their intensities increased with temperature due to the diffusion kinetics explained before. However the intensity of iron peaks decreased. This can be explained by prevention of the incident X-ray from reaching the substrate with layer thickening.

Yasavol et al. [28] investigated the effects of post-oxidizing treatment on properties of plasma nitrided AISI 4130 steel. The plasma nitriding treatment was done under pressure of 4 mbar and gas atmosphere of 75 % N_2 + 25 % H_2 at temperature of 550 °C. Post-oxidizing treatment was done under pressure of 4 mbar and at temperatures of 400 and 500 °C. Three gas mixtures were used for this treatment: 25 % O_2 + 75 % H_2, 10 % O_2 + 90 % H_2, and 100 % O_2.

After post-oxidizing treatment, the presence of several phases was reported such as Fe_3N, Fe_4N, Fe_2O_3, Fe_3O_4, and Fe. This may be due to the fact: with growing the oxide layer during oxidation treatment, the nitrogen atoms were released. One part of these released atoms which diffused into the matrix and the interface of oxide and nitride layer. Therefore, the transformation of γ′-Fe_4N to ε-Fe_3N occurred. The presence of Fe in the microstructure was attributed to two phenomena: (1) decomposition of γ′ to γ′ + (α-Fe,N) which occurs due to exposure to treatment temperature during plasma oxidizing, which produces α iron and (2) more penetration of the X-ray beam into deeper depth of the substrate because of lower absorption coefficient of iron oxides compared to that of iron nitrides.

For oxidizing atmosphere of 100 % O_2, Fe_2O_3 and Fe_3O_4 phases were detected. With adding hydrogen and using the above-mentioned gas mixtures, the amount of oxide phases decreased. Also as the oxygen content decreased in the oxidizing atmosphere, the decrement in the amount of Fe_3O_4 was more than Fe_2O_3.

The thickness of compound layer was reported to be 12 μm, while that of oxide layer was observed to be 1, 1.3, and 1.6 μm for oxidizing atmospheres of 25 % O_2 + 75 % H_2, 10 % O_2 + 90 % H_2, and 100 % O_2.

Alsaran et al. [29] also investigated the effects of post-plasma oxidizing treatment on the properties of AISI 5140 steel. Plasma nitriding was done at 500 °C, under atmosphere of 40 % N_2 + 60 % H_2 and pressure of 5 mbar for 4 h. The post-oxidizing treatment was done at 500 °C for 15, 30, and 60 min under a pressure of 2 mbar and atmosphere of only nitrogen.

With plasma oxidizing for 30 min, a 0.4 μm thick oxide layer was reported to be formed on the nitride layer of 5 μm thick. For nitrided sample, only nitride phases of γ'-Fe_4N and ε-Fe_3N as well as small amount of oxide (resulted from insufficient evacuation) were detected. For durations more than 15 min, the oxide layer made of two oxide phases of Fe_2O_3 and Fe_3O_4 was formed. In lower durations and early stages of oxidation, Fe_3O_4 was formed however with further durations, the intensities of both oxides enhanced. For duration of 60 min, α-Fe peaks also arose. This can be attributed to thinning of compound layer owing to sputtering effect and lower absorption coefficient of iron oxides.

Alsaran et al. [30] investigated the effects of post-aging treatment on the plasma nitrided AISI 4140 steel. They plasma nitrided samples with following parameters: atmosphere of 50 % N_2 + 50 % H_2, duration of 1 h, temperature of 575 °C, and pressure of 1000 Pa. The samples were quenched and then the post-aging treatment was done at 400 °C, for durations of 0.5, 1, 4, 8, and 12 h in a gas mixture of 50 % H_2 + 50 % Ar.

The microstructure of both quenched and post-aged samples consisted of compound layer and diffusion zone. As seen in Fig. 2.24, two distinct layers of γ' and ε phases are distinguishable and their separating line is obvious. The compound layer had the thickness of almost 20 μm. The quenching after nitriding resulted in the formation of α_N phase. After aging, a new needle-shaped α'' phase was initiated

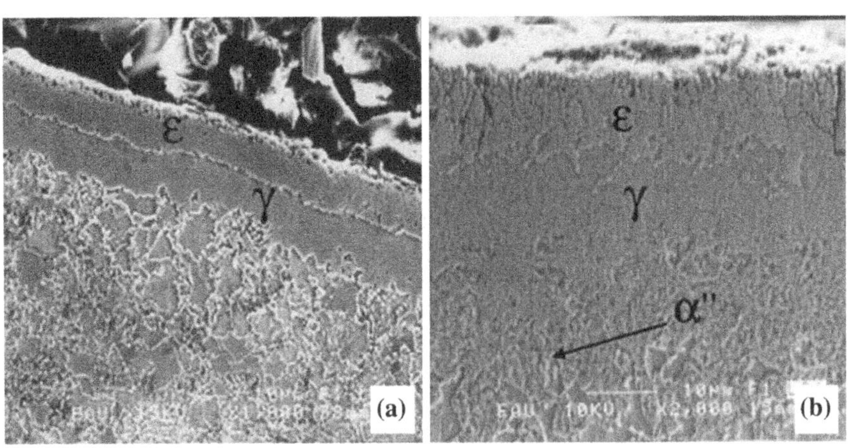

Fig. 2.24 SEM micrographs of specimens **a** quenched and **b** plasma post-aged for 1 h [30]

from martensite. The most intense α'' phase was observed for duration of 1 h because of the increasing possibility of nitrogen diffusion due to ion bombardment.

After aging treatment the amount of γ' and ε phases decreased and α phase was detected. The decomposition of nitride phases resulted in the releasing of nitrogen and its diffusion into the α phase. This supersaturated phase also was transformed to more stable α''-$Fe_{16}N_2$. The amount of this phase reached its maximum value for the duration of 1 h.

Alves et al. [31] studied the effect of three different cooling rates on the microstructural properties of the AISI 1010 steel. Plasma nitriding procedure was done with following parameters: atmosphere of 20 % N_2 + 80 % H_2, temperatures of 500 and 580 °C, duration of 2 h, and pressure of 500 Pa. Three cooling method was applied on the nitrided samples: (1) cooling naturally on the sample holder, (2) cooling on a cold surface and (3) cooling in an oil medium.

In the case of nitriding at 500 °C, two distinct layers of compound layer (of 4 μm) and diffusion zone were observed. For cooling procedures of 2 and 3, a clear boundary was not observed between diffusion zone and the substrate. For the sample cooled on the holder with total nitrided layer of 300 μm, two diffusion zones were distinguished: First, a zone nearer to the surface with the thickness of 30 μm with no nitride precipitation and second zone just after that toward the core composed of γ'-Fe_4N with small precipitates of α''-$Fe_{16}N_2$ phase formed near the substrate interface.

In the case of treatment at 580 °C, the thickness of compound layer increased to 17 μm. Moreover, for all three samples another layer of 23 μm was observed between compound zone and diffusion layer named transformation zone. This layer consisted of nitride precipitates in grain boundaries for cooling on the substrate. However no nitride precipitate was observed for two others.

The same two-part diffusion zone was also observed for this sample except the thickness of first zone extended to 110 μm. Again, no precipitate for first zone and two nitrides of γ' and α'' were detected for the first and second layer, respectively.

The compound layer of all samples (nitrided at two temperatures and cooled in three ways) was constituted by ε-$Fe_{2-3}N$ and γ'-Fe_4N. For nitriding at 500 °C, the amount of ε and γ' was 10 and 90 % for substrate cooling and 17 and 83 % for oil cooling. This refers to the fact that, γ'-Fe_4N is formed by transformation of ε-$Fe_{2-3}N$ phase and by rapid cooling this transformation is hindered and the amount of ε phase remains higher. For the samples nitrided at 580 °C, phase constituents in three depths are presented in Table 2.10. The presence of γ-Fe was attributed to the difference between the real temperature of the sample and the nominal temperature and reaching the temperature to austenitization area.

Ashrafizadeh [32] studied the plasma nitriding of Ck45 and En40B (32CrMo12) structural steels and compared with gas nitriding. Plasma nitriding was done for 5 h and gas nitriding was done for 70 h.

It was observed that the compound layer formed for gas nitriding was thicker and mainly constituted by γ' and ε phases as well as some iron oxide (Fe_3O_4) while that of plasma nitrided sample composed of just γ'. This difference in thickness was attributed to the sputtering and material removal of the surface in plasma nitriding.

Table 2.10 Phases obtained during a progressive wear in surface of nitrided samples for depths of 35 and 65 μm at temperature of 580 °C [31]

Position	Cooling	Phases
Diffusion zone 1 (depth of 65 μm)	Substrate holder	γ-Fe and α-Fe
	Oil reservoir	γ-Fe and α-Fe
Transformation one (depth of 35 μm)	Substrate holder	γ'-Fe_4N, ε-$Fe_{2-3}N$, γ-Fe and α-Fe
	Oil reservoir	γ'-Fe_4N, ε-$Fe_{2-3}N$, γ-Fe and α-Fe
Compound layer (surface)	Substrate holder	γ'-Fe_4N and ε-$Fe_{2-3}N$
	Oil reservoir	γ'-Fe_4N and ε-$Fe_{2-3}N$

For both gas and plasma nitriding and for both steels, a diffusion zone was observed below the compound layer in which nitrogen dissolves in the ferritic phase and forms nitrides. For En40B steel, alloying elements with higher affinity to nitrogen formed precipitates.

2.2 Mechanical Properties

2.2.1 Austenitic Stainless Steels

According to Wang et al. [2] (see Sect. 2.1.1), the microhardness for AISI 304L austenitic stainless steel increased with temperature with maximum value of 1200 $HV_{0.1}$ at 460 °C as shown in Fig. 2.25. This increase was probably due to the formation of chromium nitrides.

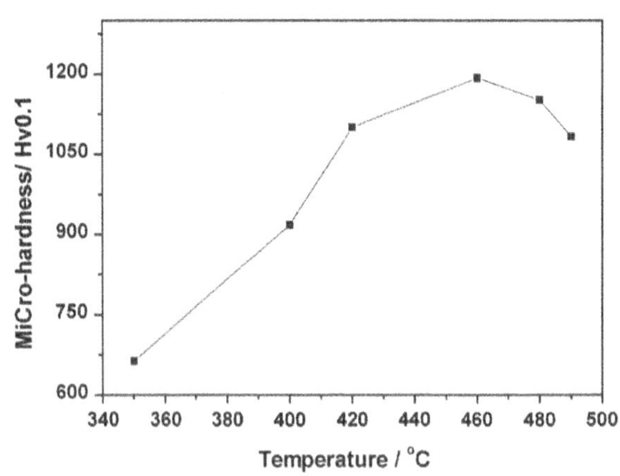

Fig. 2.25 Microhardness of AISI 304L austenitic stainless steel plasma nitrided at various temperatures [2]

They have found that, increasing nitriding temperature from 350 to 480 °C, increases the wear rate slightly, which is still much lower than that of the unnitrided material. Formation of a hard s-phase layer was the reason for achieving the lowest wear rate at 350 °C. Increasing chromium nitrides concentration in the modified layer corresponded for decreasing wear resistance with increasing temperature. The highest wear rate belongs to 480 °C treated sample where a single CrN nitride layer was produced.

According to the studies of Stinville et al. [3] (see Sect. 2.1.1) about the fatigue life of plasma nitrided AISI 316L steel, plasma nitriding led to an increase of the total stress amplitude during the whole cyclic deformation test. As seen in Fig. 2.26, this effect was not symmetric.

They estimated a residual compressive stress value of 1.7–3 GPa for plasma nitrided samples. An improvement was also seen in the number of cycles to failure as a function of the nitriding duration. This can be ascribed to the pressure of residual compressive stress. The residual compressive stress (3 GPa) and the fatigue life (5.5×10^4 cycles) reached their maximum value. In longer durations the fatigue life decreased because some grains are lost favoring an early crack initiation.

Egawa et al. [5] (see Sect. 2.1.1) also found that wear properties of JIS-SUS 304J3 stainless steel were improved by plasma nitriding. Friction coefficient of the samples (plasma nitrided at 400 °C) was almost the same against 304 and 52,100 balls. However, friction coefficient against an alumina ball was 0.55 while that of untreated specimen was 0.7.

Also they reported that the wear volume of the nitrided samples was less than one tenth of those of untreated specimens.

Li et al. [4] (see Sect. 2.1.1) did some experiments to find a relationship between surface hardness and plasma nitriding temperature for AISI 316L austenitic

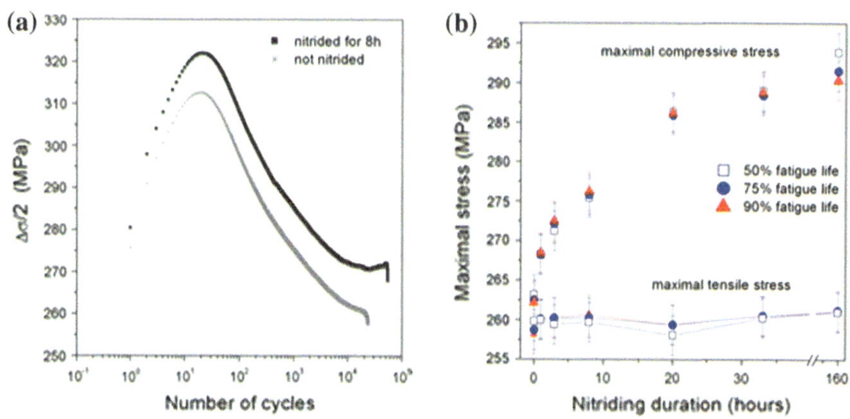

Fig. 2.26 **a** Cyclic stress amplitude evolution during fatigue tests for untreated and 8 h treated specimens for a $\Delta\varepsilon p/2 = 2 \times 10^{-3}$ plastic strain amplitude and **b** maximal compressive (*dotted line*) and tensile (*full lines*) stresses at different stages of the fatigue life for $\Delta\varepsilon p/2 = 2 \times 10^{-3}$ plastic strain amplitude [3]

Table 2.11 Wear rate of AISI 316L austenitic stainless steel plasma nitrided with different parameters [4]

Process	Process parameters	Phase composition	Surface microhardness ($HV_{0.1}$)	Wear rate (mg km^{-1})
No. 0	Unnitrided	γ (Matrix)	219	2.06
No. 1	480 °C × 600 Pa × 4 h	CrN	1117	0.98
No. 2	450 °C × 600 Pa × 4 h	CrN + S-phase	1089	0.58
No. 3	420 °C × 600 Pa × 4 h	CrN + S-phase	939	0.33
No. 4	400 °C × 600 Pa × 4 h	S-phase	763	0.21
No. 5	350 °C × 600 Pa × 4 h	S-phase	735	0.22

stainless steel. They found a very steep trend for microhardness increasing on the surface of the sample (as found by Wang et al. [2] shown in Fig. 2.25) when nitrided at temperatures below 400 °C, but it turned to a gradual trend when temperature reached beyond 450 °C. The formation of chromium nitrides corresponded for this dramatic increasing in hardness. However this can have a deleterious effect and lead to decreasing of wear resistance as reported by the authors.

As can be seen in Table 2.11, plasma nitriding improved wear resistance of steels with respect to untreated ones. The rate of wear completely depends on the treatment temperature. The lowest wear rate obtained with nitriding at 400 °C led to the formation of a hard s-phase layer corresponding for the low wear rate. In addition, increasing temperature led to the increment of chromium nitrides formation and this was ascribed to the reduction in dry sliding wear resistance. More existing chromium nitrides in the layer deteriorated wear resistance of AISI 316L steel. The dry sliding wear rate of nitrided steel reached a maximum value at 480 °C where a single CrN nitride layer was produced. Therefore, 350 and 400 °C nitriding temperatures led to the best wear properties because of maximum amount of s-phase and no chromium nitride.

Figure 2.27 shows the SEM images of the worn surfaces of both unnitrided and nitrided AISI 316L samples after dry sliding against the GCr15 at room temperature. As seen in Fig. 2.27a, wear of the unnitrided sample was so severe. As a result

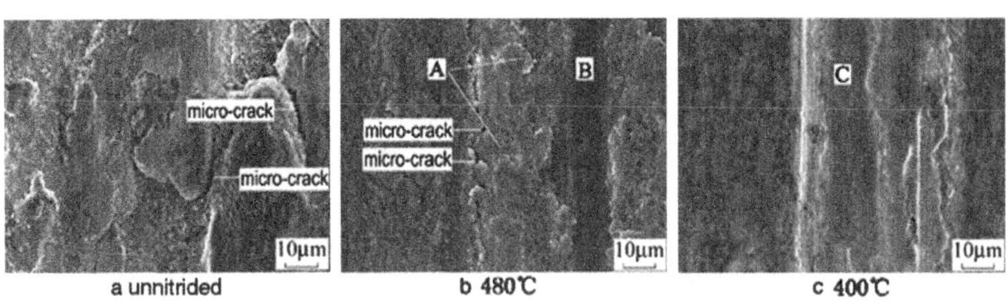

Fig. 2.27 Appearance of wear surfaces of AISI 316L austenitic stainless steel: **a** unnitrided, **b** nitrided at 480 °C and **c** nitrided at 400 °C [4]

of this severe wear, the surface became very rough and experienced microcracks, peeling off, and small wear debris distributed on the wear track. The wear process began with asperity fracture and consequently, small amounts of wear particles were produced via asperity fracture. These particles then crushed down between slider and sample and then facilitated to adhesive and abrasive wear on the unnitrided samples with increasing sliding distance. Moreover, the detaching of material from the surface of the steel, its plastic deformation between slider and sample, formation of the consequent microcracks, and peeling off the materials from the surface led to the higher wear rate.

Figure 2.27b shows a typical morphology of the worn surfaces of CrN layer formed at 480 °C. Two distinct areas are distinguishable in the worn track (indicated as "A" and "B" in this figure). The area "A" was so rough and exposed signs of plastic deformation. This can be ascribed to local flash temperature at the rubbing interface which can be high enough to soften the asperities in contact area and cause their deformation under initial high contact stress. It can be considered to be aggregates of fine wear debris generated from area "B" with increasing sliding distance. The microcracks observed were because of a CrN layer formed on the surface. Although despite these microcracks, the wear rate of this sample was much lower than that of unnitrided sample.

Figure 2.27c shows a typical morphology of the worn surfaces of s-phase layer formed at 400 °C. As seen in the figure, some shallow and wide grooves were produced along the wear direction (e.g., region "C" in Fig. 2.27c). It can be concluded that the dry sliding wear resistance of single expanded austenite layer was better than single chromium nitrides layer. This can be due to the nitrogen dissolved in austenite lattice which would increase the hardness of plasma nitrided layer. As a result, wear resistance increases due to the ductile behavior of s-phase in scratching and in abrasion.

Oliveira et al. [6] (see Sect. 2.1.1) also proved increase of surface hardness by plasma nitriding. The nitride layer obtained at 500 °C possessed higher microhardness value ($\sim$1050 HV) in comparison to the samples nitrided at 400 °C ($\sim$853 HV) and 450 °C ($\sim$950 HV), as well as a larger case depth, probably due to the formation of chromium nitrides and the high nitrogen concentration on the white layer. An abrupt fall was observed beneath the nitrides layer (20 μm). Later, N increased until the depth of 40 μm, forming the s-phase. This was corresponding for high hardness obtained. After that the N content decreased to reach the substrate.

Olzon-Dionysio et al. [7] also have done hardness measurements to study edge effect (see Sect. 2.1.1). Figure 2.28 depicts hardness values in different regions on the surface of the nitrided samples (regions 1–6 on the samples B, C, and D according to Fig. 2.7). The mean hardness value for each region is represented by a continuous line. The indent penetration during hardness measurements was between approximately 0.66 μm for region 4 and 1.3 μm for region 2.

Hardness reached a maximum value in regions 1 and 4 and then decreased abruptly in regions 2 and 5. There was a small interface region toward the center of

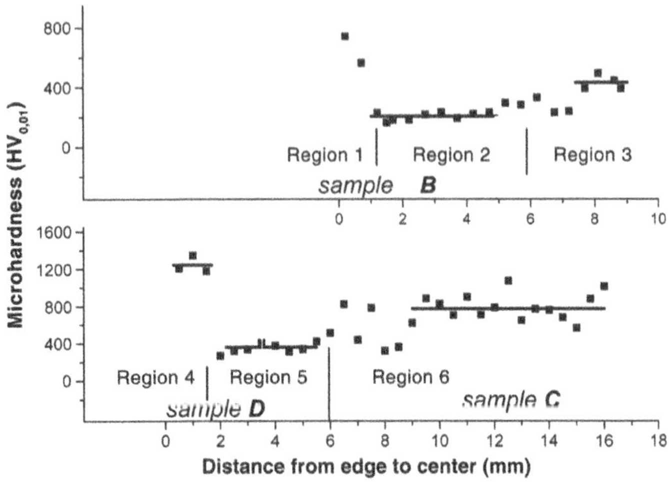

Fig. 2.28 Surface hardness values in different regions of the nitrided samples B, C, and D [7]

the sample after which hardness increased again and tended to stabilize in central regions 3 and 6.

The estimated extension of regions marked in Fig. 2.7 is presented in Table 2.12. As can be seen, despite their different diameters, the extension of equivalent regions in samples B (1 and 4) and D (2 and 5) agrees to be better than 98 %.

According to Table 2.12, the rings formed in samples B and D had the same characteristics. This shows that the formation of the rings does not seem to depend on sample geometry, at least not in their current dimensions.

It should be noted that these hardness results were in agreement with XRD results. For example, the existence of Cr_2N phase (with hardness of more than 1479 HV) on sample B can be the reason for the highest hardness values on the ring region.

Nakajima et al. [8] (see Sect. 2.1.1) evaluated the mechanical properties of AISI 304 austenitic stainless steel samples after post-solution treatment.

The surface hardness of the nitrided sample was reported to be 1360 HV with a rapid decrease to hardness of the untreated core of 153 HV within 0.1 mm. For post-solution treated sample the hardness was 280 HV, however, with higher core hardness compared to the just nitrided one. The hardness of the sample solution treated at 1200 °C for 45 min reached to the hardness of untreated sample in depth of 1 mm. This trend was also reported for durations of 60 and 90 min. Although the interior hardness for these two samples was higher compared to that of 45 min. It can be due to the resolving of the nitrides and diffusion of nitrogen to deeper distances.

Table 2.12 Measurements of region extension and mean hardness values for samples B, C, and D [7]

Position	Border or ring		Intermediate		Central	
Sample	B	D	B	D	B	C
Region	1	4	2	5	3	6
Extension (mm)	1.31 ± 0.06	1.33 ± 0.05	4.72 ± 0.08	4.63 ± 0.07	3.82 ± 0.09	9.85 ± 0.08
Mean hardness ($HV_{0.01}$)	652 ± 70	1244 ± 90	206 ± 20	360 ± 30	431 ± 40	775 ± 80

Table 2.13 Mechanical properties of unnitrided, nitrided, and post-solution treated specimens [8]

Specimen	0.2 % proof stress $\sigma_{0.2}$ (MPa)	Tensile strength σ_B (MPa)	Elongation δ (%)	Reduction of area φ (%)
Untreated	268	606	62	83
Nitrided	263	616	65	77
Nitrided and post-solution treated	282	640	62	78

Table 2.13 presents the mechanical properties of the samples. As can be seen, tensile properties were improved after post-solution treatment. This can be due to the fact that, the compound layer formed on the plasma nitrided sample was very hard, thin, and brittle. Therefore, it cannot reply well to tensile tests and consequently deteriorates the mechanical properties.

In laboratory air, fatigue limits of the untreated, nitrided, and post-solution treated samples were 290, 380, and 340 MPa, respectively. Due to the higher surface hardness of the nitrided sample, it had the highest fatigue limit among the samples.

In 3 % NaCl solution, fatigue limit was reported to remain the same as for laboratory air condition. However, for nitrided and post-solution-treated samples, the fatigue limit was reported to decrease due to the sensitization phenomenon and formation of chromium depleted zone. However, the fatigue life of the post-solution treated sample was obvious in comparison with unnitrided sample. The mechanisms for this increase in fatigue strength were reported as follows: the precipitation of chromium nitride (CrN), the solid solution of nitrogen, and slower crack growth rates in this sample compared to untreated sample. A fine precipitation of nitride was reported to correspond for this behavior.

Wang et al. [9] (see Sect. 2.1.1) have the effect of nitriding duration on the hardness of nitrided AISI 304.

It was reported that, surface hardness increased by nitriding time. This was attributed to thickening of the nitride layer and reducing the substrate effect on the surface hardness. For treatment time of 5 h, the hardness of the sample increased by approximately a factor of 5 compared with the unnitrided sample. The extremely high microhardness values observed were attributed to the large compressive stresses formed in the modified layers after nitriding.

Sun et al. [10] (see Sect. 2.1.1) studied the effects of carbon addition to the nitriding atmosphere on properties of AISI 321. Figure 2.29 shows the comparison between hardness profiles of samples treated with only nitriding and nitriding with CH_4 addition. As seen, the only nitrided sample showed a sudden decrease of hardness in the interface. However, the hardness drop became slight with carbon addition and this is an advantage of this process.

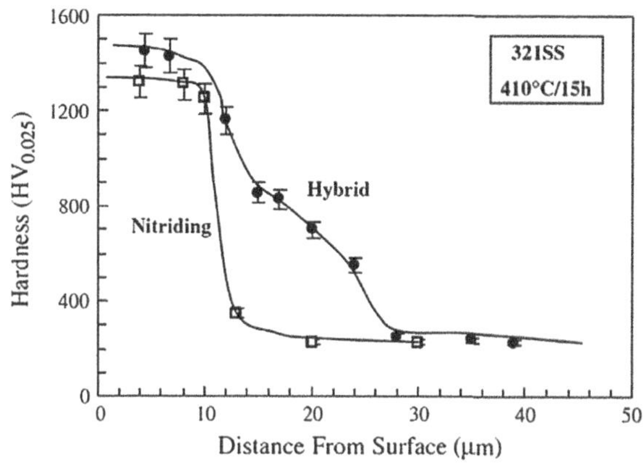

Fig. 2.29 Hardness profiles across the treated layers produced by nitriding alone and nitriding with carbon addition at 410 °C for 15 h [10]

2.2.2 *Martensitic Stainless Steels*

Alphonsa et al. [12] (see Sect. 2.1.2) also showed the increasing of hardness by nitrogen content in AISI 420 martensitic stainless steel. As can be seen in Fig. 2.30, the microhardness was maximum up to the depth of ~60 μm, at which the nitrogen concentration reduced very fast. The reduced nitrogen content resulted in a reduction in surface hardness, and the hardness value reached that of the base material by about 80–90 μm micrometers. This figure clearly shows the direct relation between nitrogen concentration and hardness in the case depth.

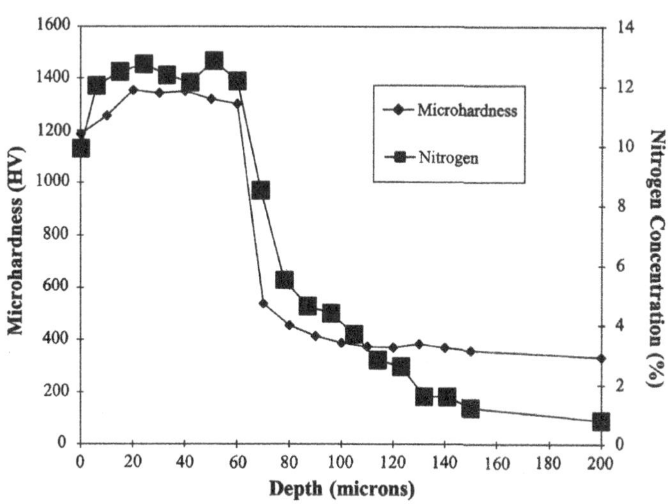

Fig. 2.30 Nitrogen concentration and microhardness values across the cross section of plasma nitrided AISI 420 [12]

Li et al. [13] (see Sect. 2.1.2) drew a conclusion about the relationship between hardness and treatment temperature for AISI 410. The increase in surface hardness was higher for the lower temperature nitrided steel. The surface hardness was 262 $HV_{0.1}$ for untreated sample and 1210, 1204, and 1113 $HV_{0.1}$ for samples nitrided at 420, 460, and 500 °C, respectively. The hardness dropped quickly and reached to the substrate value in the diffusion layer. This is suggesting that hardening in the diffusion zone was small. The effective case depths (with hardness of 350 $HK_{0.1}$) were 77, 166, and 186 μm, respectively, for the samples nitrided at 420, 460, and 500 °C.

Pinedo et al. [14] (see Sect. 2.1.2) observed the nitriding beneficial effect on hardness improvement resulted by precipitation of fine and homogeneous nitrides (γ'-Fe_4N, ε-$Fe_{2-3}N$ and CrN) on diffusion zone (Fig. 2.31). As seen in this figure, the hardness profiles experienced a sharp decrease after the maximum hardness plateau. The shape of nitrogen profile was coincident with hardness profiles across the nitrided case.

Sobiecki et al. [17] (see Sect. 2.1.2) concluded that for H9S2 steel, plasma nitriding above 600 °C was the least favorable. The smallest microhardness values were 5.39 and 6.57 GPa obtained at these temperatures.

Wu et al. [15] (see Sect. 2.1.2) evaluated the effects of two new plasma nitriding methods on mechanical properties of AISI 420. The methods were cyclic plasma oxynitriding and cyclic plasma nitriding catalyzed by rare earth La.

The microhardness of the untreated sample was 285 $HV_{0.05}$, while that of the conventional plasma nitrided sample was 920 $HV_{0.05}$. The surface microhardness of the samples treated using cyclic plasma oxynitriding and cyclic plasma nitriding catalyzed by rare earth La in four cycles, were 942 $HV_{0.05}$ and 957 $HV_{0.05}$ respectively. The higher surface microhardness resulted by these two methods was attributed to higher concentration of nitrogen on the surface. The hardness drop along the cross section resulted by conventional DC plasma nitriding was abrupt in contrast to two new methods applied. Also a higher hardness and higher nitrogen diffusion depth were observed by increasing the treatment cycles.

As can be seen in Fig. 2.32, wear behavior of the samples nitrided with three methods was similar in initial stages until 120 min but it was changed beyond this

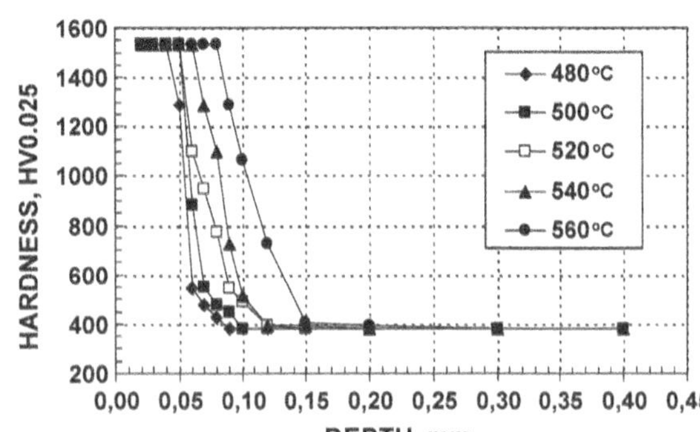

Fig. 2.31 Hardness profiles obtained for each nitriding temperature [14]

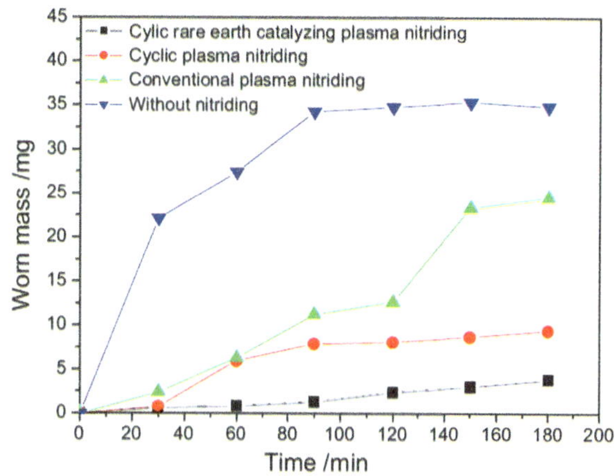

Fig. 2.32 Weight loss curves of nitrided and unnitrided AISI 420 martensitic stainless steel [15]

time. The weight loss after 180 min was 34.9, 24.6, and 9.5 mg for the untreated sample, conventional plasma nitrided sample and cyclic plasma oxynitrided sample, respectively. However, the weight loss for the sample treated by cyclic plasma nitriding catalyzed by rare earth La was 3.8 mg probably due to the higher nitrogen diffusion depth. Further densification of the surface of the nitrided case and the generation of small nitrides and consequently inducing high compressive residual stress and improving the strength, and plasticity of the nitrided layers were also counted as La functions.

Marchev et al. [33] (see Sect. 2.1.2) investigated the wear properties of AISI 410 martensitic stainless steel nitrided at different temperatures for different durations.

Plasma nitriding treatments were done at 400–566 °C in atmosphere of 65 % N_2 + 35 % H_2 under pressure of 2.7–2.8 torr for durations between 6 and 120 h.

It was concluded that lower nitriding temperatures as well as longer durations led to higher wear resistance and lower friction coefficient when compared to higher temperatures treated samples as well as untreated sample. This can be seen from the obtained data presented in Table 2.14.

2.2.3 *Precipitation Hardening Stainless Steels*

According to Esfandiari et al. [16] (see Sect. 2.1.3) for A286 (AISI 600) stainless steel, at low temperatures (≤420 °C) the surface hardness increased by time. However, the surface hardness of the high-temperature (460 and 500 °C) treated samples initially increased by time but after a maximum point, decreased.

Plasma nitriding at low temperatures led to the formation of s-phase. The hardness of this phase is very high which is due to the solid solution hardening mechanism. However, with increasing the nitriding temperature up to 460 °C (the critical temperature for the formation of chromium nitrides) the surface hardness of

Table 2.14 Results from friction and wear experiments on nitrided AISI 410 martensitic stainless steel samples [33]

Treatment temperature (°C)	Treatment time (h)	Number of rotations	Wear depth (μm)	Appearance of wear surface	Friction coefficient (μ_{avg})
None	None	24,000	17.1	Adhesion and pitting	0.7–0.8
400	96	50,000	13.7	–	0.2
400	120	50,000	10.2	Mild abrasion	Lower than 0.1
455	96	50,000	28.6	Abrasion and plastic deformation	0.4–0.5
465	96	50,000	24.8	Plastic deformation and platelet formation: batl fracture	0.4–0.5

nitrided samples increased. This effective hardness increment can be ascribed to the precipitation of fine chromium nitride particles as well as solid solution. However, with further increasing of treatment temperature to 500 °C, the decreased effect of solid solution hardening effect reduced the surface hardness.

After dry sliding wear test, the wear volume loss was very high for untreated samples. The wear volume loss was 0.6 mm^3 for untreated sample and 0.01, 0.004, and 0.001 mm^3 for samples treated at 350, 420, and 500 °C. Also, the wide and deep wear tracks were visible after test indicating poor wear resistance. However, wear tracks on the plasma nitrided samples were shallow and superficial. Increasing treatment time resulted in improving wear resistance because of thickening of the layer.

Wear tests revealed the fact that for untreated samples, the wear mechanism was adhesive. However, in plasma nitrided samples, wear occurred in mild micro-abrasion and oxidative wear modes, in contrast to the severe adhesive and abrasive wear.

Increasing hardness with temperature was also reported by Li et al. [18] (see Sect. 2.1.3). The surface hardness varied between 950 $HV_{0.1}$ and 1270 $HV_{0.1}$ for samples treated at temperatures of about 350 and 480 °C, respectively. The maximum hardness value was 1270 $HV_{0.1}$ which was measured from the surface of sample nitrided at 480 °C. This was about 3.5 times as hard as the untreated sample (362 $HV_{0.1}$). Due to close relation of microstructure with the properties, the formation of chromium nitrides caused an abrupt increase in hardness at temperature 420 °C.

They also investigated wear behavior of this steel. It was found that, the treatment temperature affected surface roughness after plasma nitriding (Table 2.15). The interaction between ions and the sample surface through the superficial sputtering (existing on the cathode surface) corresponded for this roughness variation. An increasing treatment temperature resulted in increasing energy of incident species and consequently rougher surface. Therefore, increasing temperature more

Table 2.15 Wear rate of 17-4PH stainless steel nitrided at different parameters [18]

Process parameters	No. 0 unnitrided	No. 1 480 °C/600 Pa/4 h	No. 2 450 °C/600 Pa/4 h	No. 3 420 °C/600 Pa/4 h	No. 4 400 °C/600 Pa/4 h	No. 5 350 °C/600 Pa/4 h
Phase composition	α (Matrix)	CrN + α	CrN + α	CrN + α'_N	S-phase + α'_N	S-phase + α'_N
Microhardness ($HV_{0.1}$)	362	1278	1187	1140	1022	973
Roughness R_a (μm)	0.05	0.25	0.21	0.18	0.11	0.10
Roughness R_{max} (μm)	0.1	2.4	2.3	2.1	1.32	1.12
Wear rate (mg km^{-1})	10.35	1.94	1.85	1.49	0.85	1.27

than 400 °C can have deleterious effect on the wear behavior of the plasma nitrided sample.

According to Table 2.15, the highest dry sliding wear resistance was obtained at treatment temperature of 400 °C. The wear resistance of the 400 °C/4 h treated sample was higher than the 350 °C/4 h treated one since the layer formed at 400 °C was thicker and harder in comparison to layer formed at 350 °C. The other important factor in good wear resistance was the formation of s-phase.

The wear of the untreated 17-4PH is dominantly characterized by strong adhesion, abrasion, and oxidation mechanism (as mentioned before).

Existence of microcracks and flake-like debris with slight plastic deformation on the worn surface of the 480 °C/4 h treated sample, indicated occurring of a mild adhesive and abrasive wear. Therefore, dry sliding wear resistance of nitrided samples was improved. The surface roughness of treated sample reduced the contact area compared to the untreated one and consequently increased the pressure on the asperities. Therefore flattening of the asperities occurred as a result of greater local plastic deformation. Consequently the contact area increased and reduced the contact pressure, leading to a mild adhesive wear and abrasive wear mechanism. Although the surface roughness was higher compared to the unnitrided surface, its wear rate was lower than that of the unnitrided samples. This was due to the formation of a hard and rigid chromium nitride (CrN) layer on the 480 °C/4 h treated sample surface. The oxidative film can play the role of lubricant during the test and reduce the friction between slider and sample.

The worn surface of the 400 °C/4 h treated samples showed smooth worn surface without microcracks. For the 400 °C/4 h, treated layer was composed of s-phase and α_N. Due to the ductile behavior of s-phase in scratching and abrasion, the formation of s-phase during plasma nitriding improved the wear resistance. The surface roughness of this sample increased the contact area with GCr15 even more, compared to the 480 °C/4 h treated sample. This led to the reduction of initial asperity deformation and polishing. Therefore, dry sliding wear resistance of the 400 °C/4 h treated sample was improved compared to that of the 480 °C/4 h treated sample.

2.2.4 *Tool Steels*

Hardness measurements done by Birol [20] (see Sect. 2.1.4) in case of plasma nitrided X32CrMoV33 steel confirmed the microstructure observations mentioned in Sect. 2.1.4. As shown in Fig. 2.33, hardness decreased sharply from surface to the core. The highest hardness belonged to the surface due to the nitrides precipitation and to the supersaturation with nitrogen of the BCC matrix in the diffusion zone. The case depth which was estimated from the hardness profile to be approximately 150 μm, was consistent with the etching response of the surface layer. The atomic Fe:N ratio in the compound layer was close to 3. This ratio decreased through the substrate.

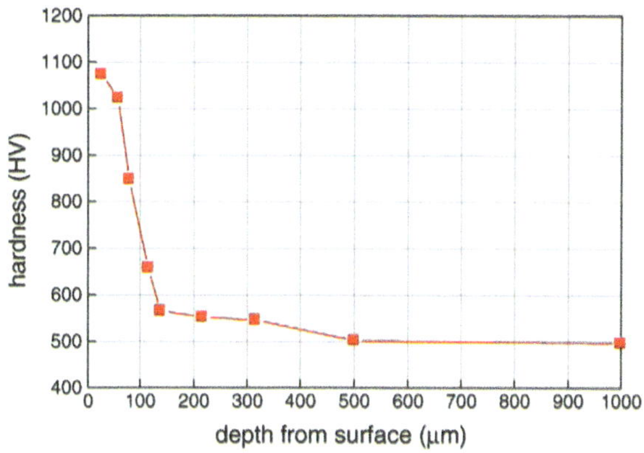

Fig. 2.33 Hardness variations with depth from the surface of the plasma nitrided hot work tool steel sample [20]

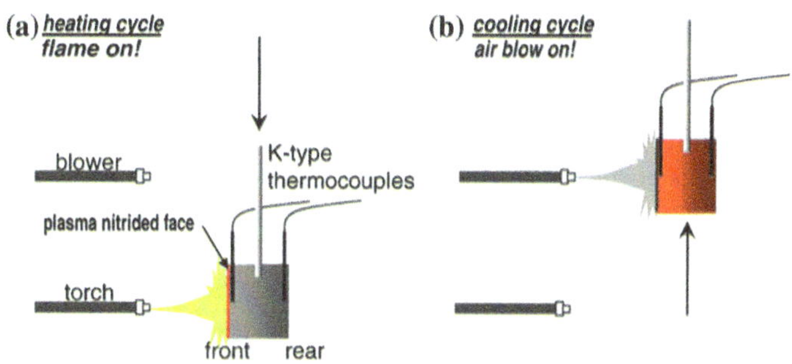

Fig. 2.34 Schematic of thermal fatigue test setup for the: **a** heating and **b** cooling cycles [20]

Thermal fatigue test was done to approximate thermal loading conditions. Plasma nitrided samples were cycled between 750 and 450 °C. The experimental setup for this test is presented in Fig. 2.34.

The thermal fatigue life of the nitrided tool steel obtained by test was much shorter than expected. Therefore, plasma nitriding resulted in a decrease in thermal fatigue life.

The first oxide to form on the surface of the steel is Fe_3O_4 and with further oxidation transforms into Fe_2O_3. The failure of the oxide scale was enhanced because of the poor adherence and limited ductility of these oxides. Reduction of adhesive strength of the oxide film on the substrate occurred due to the thermal expansion mismatch with the underlying metal and the consequent compressive stresses. Isotherm annealing (at 750 °C for 6 h) retained the oxide scales on the surface.

This oxidation and removal and reoxidation resulted in a dimple profile on the front face. The resistance to oxidation was improved by plasma nitriding. This can

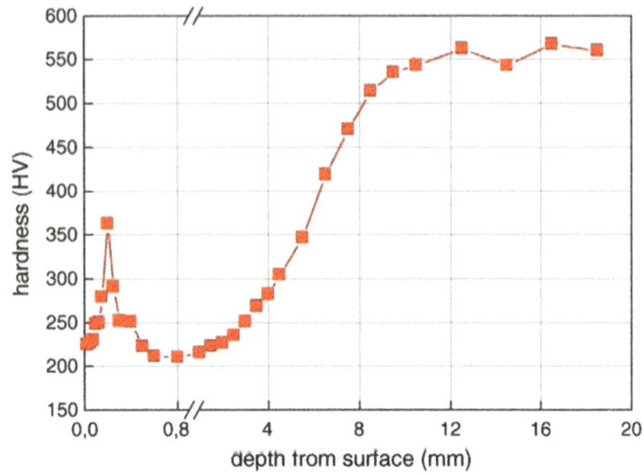

Fig. 2.35 Hardness variations versus depth from the surface of the plasma nitrided hot work tool steel sample after 500 thermal cycles [20]

be confirmed by formation of a thin Fe_3O_4 layer on the nitrided surface after isothermal annealing at 750 °C for 1 h under a vacuum of 10^{-3} Pa. Therefore, the oxidation of nitrided tool steel occurred more easily than the hot work tool steel, probably due to the fast-diffusion paths which were created by population of defects in the surface.

According to Fig. 2.35, the surface hardening caused by plasma nitriding was completely eliminated during thermal fatigue. The tempered martensitic structure was replaced by fine, equiaxed ferritic grains implying a dynamic recrystallization process during thermal cycling. This was ascribed to thermal stresses generated at the surface during thermal fatigue test.

Nouveau et al. [21] (see Sect. 2.1.4) have observed improving mechanical properties of 90CrMoV8 tool steel by plasma nitriding.

At constant temperature of 500 °C, hardness of subsurface was measured to be around 1250 (10 and 12 h) or 1050 (6 and 8 h) HV while the hardness of bulk was 650 HV. With increasing the treatment duration from 6 to 12 h, the hardened layer thickness varied from approximately 40–120 μm.

The nitrogen content severely decreased from surface toward the bulk. Nitriding at 500 °C for 10 h led to the highest nitrogen content and the hardest layer.

At 460 °C, a very thin hardened layer (40 μm thick) was obtained with a hardness of around 1150 HV (Fig. 2.36b). At 460 °C, the thickness of the diffusion layer was too low for withstanding the severe abrasion occurring in wood machining. As Fig. 2.36 shows, the diffusion layer consisted of two parts: compound layer as outer part and hardened bulk material as inner part. The top surface of 520 °C nitrided sample was less hard than the one nitrided at 500 °C and above a 100 μm depth the microhardness reduced to 300 HV.

For above-mentioned reasons, the optimal nitriding parameters were selected as temperature of 500 °C and duration of 10 h.

Pessin et al. [23] (see Sect. 2.1.4) observed the significant improvement in the wear resistance after plasma nitriding for AISI M2 tool steel. The obtained results are shown in Table 2.16.

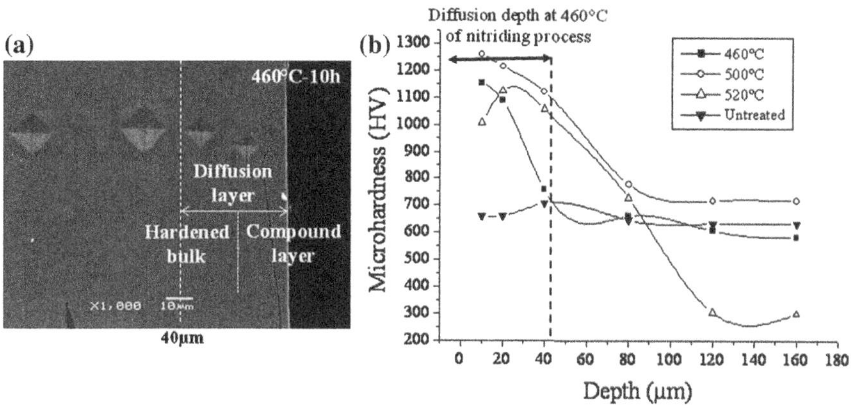

Fig. 2.36 **a** SEM cross-section micrograph for the steel nitrided at 460 °C for 10 h, and **b** hardness variations with depth for steels nitrided at different temperatures for 10 h [21]

Table 2.16 Surface hardness, friction coefficient, scar length, and wear depth [23]

	Temperature					
	450 °C			500 °C		
	200 Pa	400 Pa	900 Pa	200 Pa	400 Pa	900 Pa
Surface hardness ($HV_{0.2}$)	1178	1274	1263	1260	1286	1272
Friction coefficient	0.80	0.25	0.25	0.80	0.25	0.25
Scar length (mm)	2.9	1.3	1.8	2.5	1.3	1.3
Wear depth (μm)	42.1	8.5	16.2	31.3	8.5	8.5

The high friction coefficient ($\mu = 0.95$) caused significant adhesion on the surface of the untreated specimen. The existence of a thin compound layer significantly reduced the adhesive wear and friction coefficient ($\mu = 0{:}25$) and consequently improved wear resistance. The samples plasma nitrided at 200 Pa showed spalling during the wear test, which was believed to be due to the lack of a compound layer. The high friction coefficient measured on the diffusion zone ($\mu = 0{:}80$) resulted in high shear stresses on the sample surface, which combined with its low toughness produces a significant material loss. Due to the hardness increase from nitride precipitation in the diffusion zone, the plastic deformation capability of the material was reduced, resulting in a brittle fracture mechanism.

2.2.5 *Structural and Constructional Steels*

Şengül et al. [34] studied the effects of plasma nitriding on fatigue crack growth on AISI 4140 steel under variable amplitude loading. They used these parameters for

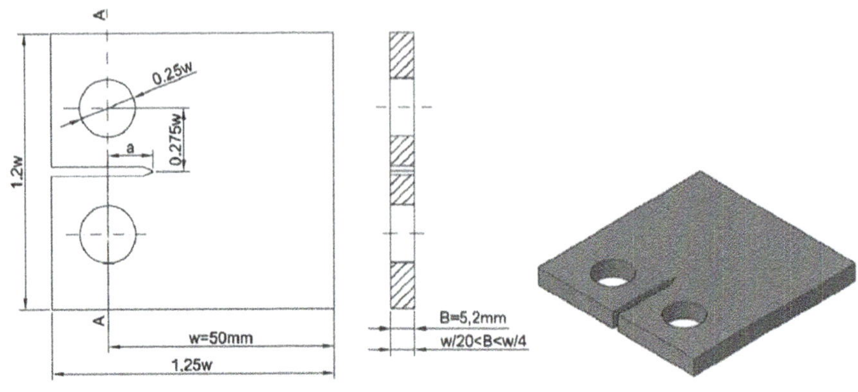

Fig. 2.37 Dimensions of the CT specimen used for testing [34]

nitriding: temperature of 500 °C, duration of 2 h, atmosphere of 50 % N_2 + 50 % H_2, and pressure of 500 Pa. The compact tension (CT) specimens (Fig. 2.37) were fabricated in L–T orientation, i.e., the load was applied in the longitudinal direction and the crack was in the transverse direction. Fatigue crack growth (FCG) tests were done in two conditions to evaluate fatigue life of samples: Constant amplitude loading (CAL) and Variable amplitude loading (VAL). Constant amplitude loading was defined as fatigue under cyclic loading with constant amplitude and a constant mean load. Much larger retardations were observable in tests with a high peak load added to constant amplitude loading. Such high loads were frequently called overloads. If constant amplitude loading includes one or more overload cycles, then this is called variable amplitude loading.

For the nitrided sample, the maximum hardness was achieved on the surface. The substrate hardness in the unnitrided condition was measured at 200 $HV_{0.025}$.

Plasma nitriding treatment improved the fatigue crack growth limit under constant amplitude loading (Fig. 2.38a, b). These improvements were the result of the hardening of the material surface in the plasma nitriding case. The hardened surface caused by the plasma nitriding treatment increased resistance to the fatigue crack growth rate.

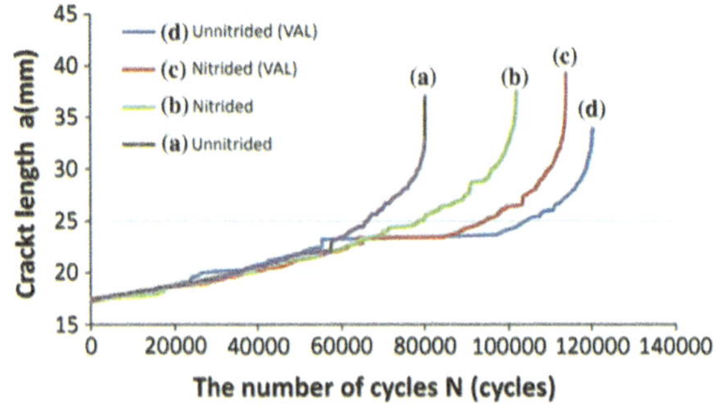

Fig. 2.38 Crack development diagram of compact samples of 4140 steel: **a** unnitrided, **b** nitrided, **c** nitrided with 10 cycles overload, and **d** unnitrided with 10 cycles overload [34]

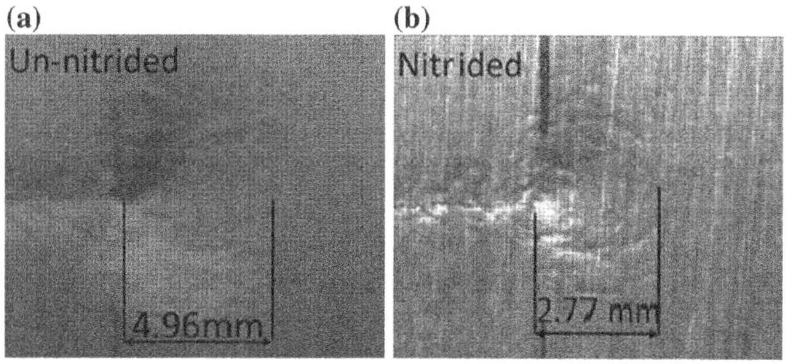

Fig. 2.39 The surface images of the plastic zone after overloading for: **a** unnitrided and **b** nitrided sample [34]

For unnitrided specimens under the variable amplitude loading condition (during the retardation period), the crack growth significantly decreased. On the other hand, the plasma nitriding process under the variable amplitude loading condition decreased the life of fatigue crack growth (Fig. 2.38c).

According to Fig. 2.39 after overloading, the surface of the plastic zone for unnitrided sample was greater than nitrided one. During overloading, the hard nitride layer resisted surface movement and consequently swelled and peeled off. This resistance prevented more growth of plastic zone, therefore, a shorter retardation period was caused by overloading.

Li et al. [24] (see Sect. 2.1.5) investigated fretting fatigue properties of plasma nitrided 709M40 steel.

They found that the average thickness of the compound layer was increased from 4.3 to 7.5 μm when the nitrogen content in the atmosphere increased from 25 to 65 %. The average hardness of the compound layer (surface hardness) was 840 $HV_{0.1}$. The residual stress in the nitrided case was compressive and its magnitude gradually reduced with increasing distance from the surface.

As shown in Fig. 2.40, plasma nitriding significantly improved fretting fatigue strengths. The improvements were ascribed to the existence of compound layer,

Fig. 2.40 Fretting fatigue S–N curves of unnitrided and plasma nitrided 709M40 samples. Plasma nitriding was performed at 500 °C for 10 h, with nitrogen potentials of 25 and 65 % [24]

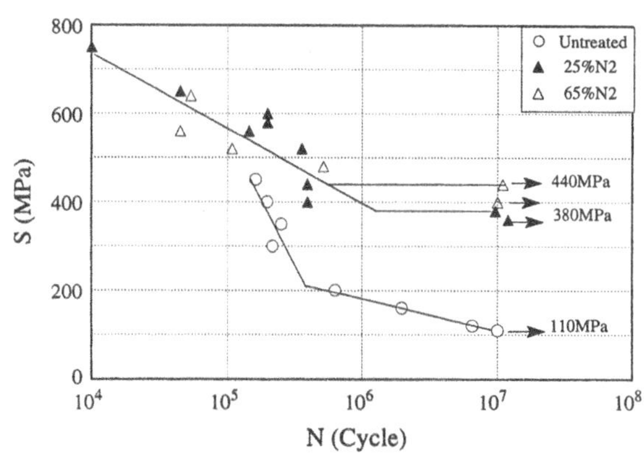

diffusion layer, compressive residual stress, and also friction coefficient of the plasma nitrided steel.

Comparing fretting fatigue S-N curves in Fig. 2.40 shows the effect of compound layer. This change in fretting fatigue strength can only be attributed to the different compound layer produced at two nitrogen potentials. Probably, the high hardness of the compound layer increased the resistance to fretting surface damage and surface crack initiation. The thicker compound layer acted as a thicker hard coating and provided more protection against fretting. As a result, the fretting fatigue strength improved.

Compared with traditional nitriding methods, plasma nitriding produced much more ductile compound layer due to the sputtering effect during the plasma nitriding.

Increasing nitriding time led to an increase in the case hardness and case depth. Nitriding durations of 3, 10, and 40 h at 500 °C produced case depths of 187, 340, and 552 μm, respectively. The hardest diffusion zone was achieved by the longer nitriding duration of 40 h.

The fretting fatigue strengths (10^7 cycles) of the nitrided samples were similar. This can be proved by Fig. 2.41 (10 and 40 h treated samples). As for the 3 h nitrided sample, despite significant difference with others in case depth, there was little difference in fretting fatigue strength. It can be concluded that, case depth and a high case hardness in the diffusion zone were not major factors in determining the fretting fatigue strength.

Introducing compressive residual stresses led to improvements in fatigue life. In contrast, degradation was caused by introducing tensile residual stresses.

Table 2.17 summarizes the surface hardness, the surface residual stress and the case depth of the nitrided samples. Nitriding for 3 h produced a high compound layer hardness (794 $HV_{0.1}$) and a high surface compressive residual stress (−502 MPa). The combination of these properties significantly increased the fretting fatigue strength (10^7 cycles) of the steel from 110 to 340 MPa. However, since the compound layer was thin (3.1 μm), and the compressive residual stress distribution was only to a very shallow depth, once cracks are initiated at high stresses

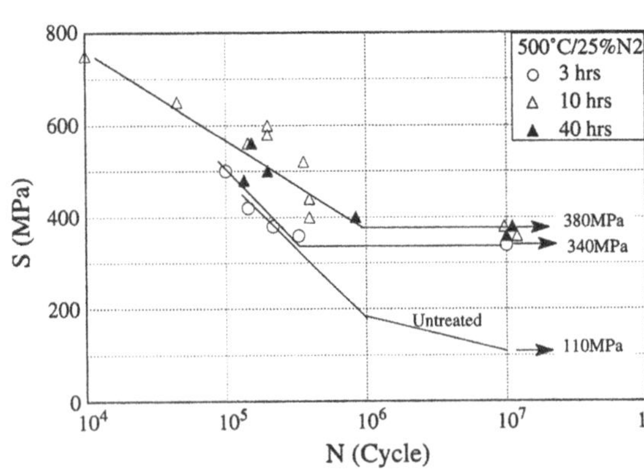

Fig. 2.41 Fretting fatigue S-N curves of 709M40 samples plasma nitrided at 500 °C for 3, 10, and 40 h with nitrogen potential of 25 % [24]

Table 2.17 Influence of nitriding time (at 500 °C in 25 % N_2:75 % H_2) on properties of nitrided samples [24]

Nitriding time (h)	3	10	40
Surface hardness, $HV_{0.1}$	794	845	789
Compound layer thickness (μm)	3.1	4.7	6.3
Surface residual stress (MPa)	−502	−498	−385
Case depth (μm)	187	340	552

from the contact surfaces, the hard compound layer and the shallow compressive residual stress depth could not have any further influence on the propagation of the cracks. Therefore, it can be concluded that to improve the fretting fatigue behavior of nitrided samples, a hard and relatively thick compound layer and a high and deep compressive residual stress are required. However, longer duration and deep case depth are not necessary.

Karaoğlu [25] (see Sect. 2.1.5) studied the effects of plasma nitriding on the mechanical properties of AISI 5140 steel. According to Table 2.9, longer durations at 500 °C or performing the treatment at 550 °C led to decrease in hardness.

Among the treated samples, the highest initial wear rate after 400 m sliding and maximum total wear after 200 m sliding belonged to 455,020 and 150,020 samples, respectively.

The charts of friction coefficient (μ) versus sliding distance showed that at the steady-state period, the friction coefficient values of plasma nitrided samples were slightly greater than the untreated samples. The maximum resistance to wear was shown by the maximum hardness. This can reveal the importance of surface hardness with respect to wear rate.

The thick compound layer (composed of $\gamma' + \varepsilon$) produced on the high-temperature (550 °C) treated samples was extremely brittle and increased the wear rate of the material. The particles generated from the breakdown of the thick compound layers at the beginning of the wear tests changed the wear mechanism from adhesive to abrasive. Therefore, the wear rate increased with thickening the compound layer.

The large and predominantly metallic debris particles from the unnitrided sample (Fig. 2.42a) were in line with the severe wear experienced by that sample. However, the wear debris produced from the nitrided sample was a mixture of plate-like large products of early periods and very fine brown powders (Fig. 2.42b). The oxidized appearance of the worn surface and the generation of oxide wear debris indicated the dominancy of mild oxidative wear in the wear of the nitrided samples.

Karakan et al. [26] (see Sect. 2.1.5) studied hardness changes with different gas mixtures (Table 2.18). They found that maximum hardness was in the range of 760–770 $HV_{0.05}$ for the atmosphere of 10 % N_2 + 90 % H_2.

According to Mahboubi et al. [27] (see Sect. 2.1.5), nitriding at low temperature (350 °C), lowered the nitrogen uptake and precipitate (nitride) density and consequently the hardening effect. With increasing the nitriding temperature to 550 °C,

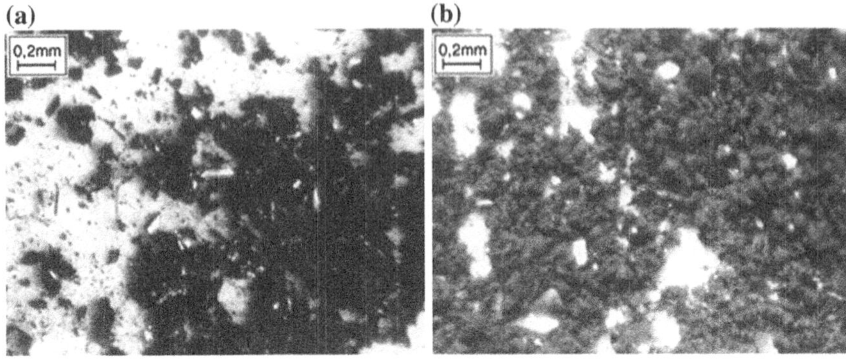

Fig. 2.42 Morphology of wear debris particles for **a** unnitrided and **b** plasma nitrided AISI 5140 (in a 20 % N_2 + 80 % H_2 gas mixture for 4 h at 500 °C) [25]

Table 2.18 Variations in the compound layer thickness, the case depth, and surface hardness with gas mixtures [26]

Plasma nitriding parameters			Compound layer (μm)	Case depth (μm)	Surface hardness (HV)
Time (h)	Temperature (°C)	Gas mixture			
4	450	90 % N_2 + 10 % H_2	4–5	160–190	540–550
4	450	10 % N_2 + 90 % H_2	2–2.5	210–240	760–770
4	450	90 % N_2 + 10 % Ar	–	35–65	290–300
4	450	10 % N_2 + 90 % Ar	8–10	60–90	470–480
4	450	80 % N_2 + 10 % H_2 + 10 % Ar	5–6	110–140	470–480
4	450	10 % N_2 + 10 % H_2 + 80 % Ar	5.5–6.5	90–120	500–510
4	450	10 % N_2 + 80 % H_2 + 10 % Ar	2–3	185–215	700–710

both precipitate growth and tempering of the matrix significantly enhanced and the precipitates got larger in size and lower in number leading to lower hardness. Therefore, it was observed that plasma nitriding of DIN 1.6959 steel at 500 °C produced the optimum nitride precipitate size and density, and resulted in the highest surface hardness.

Increasing the treatment temperature increased the roughness of the surface due to the severe sputtering of the substrate caused by nitrogen ions and higher applied voltage required for the treatment. As a result, the kinetic energy of the incident ions (or atoms) increased (Fig. 2.43). At low treatment temperatures, the formation of nitrides resulted in volumetric expansion and surface relief and consequently increment of the surface roughness.

Yasavol et al. [28] (see Sect. 2.1.5) investigated the effects of post-oxidizing treatment under different atmospheres on mechanical properties of 4130 steel. Plasma nitriding treatment improved the surface hardness from 230 $HV_{0.05}$ for

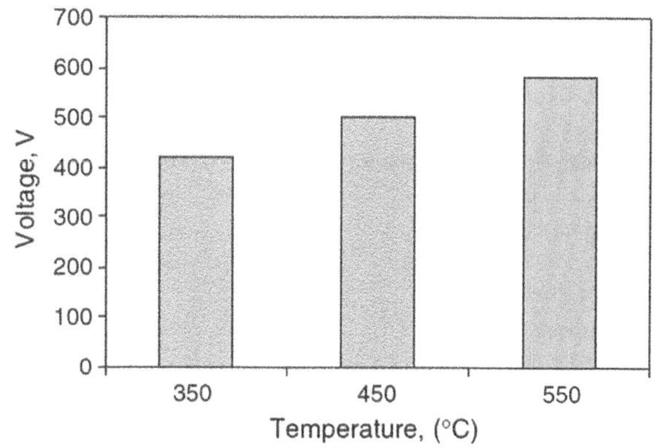

Fig. 2.43 Applied voltage at different plasma nitriding temperatures [27]

untreated sample to 870 $HV_{0.05}$ for plasma nitrided sample. This improvement was owing to the formation of γ'-Fe_4N nitride phase. However, the following oxidizing treatment decreased the improved surface hardness due to the lower hardness of oxide layer compared to nitride layer. The hardness of the layer formed under atmosphere of 25 % O_2 + 75 % H_2, was higher compared to two others due to two reasons: (1) the higher hardness of Fe_3O_4 phase and (2) the interface of this oxide film that is denser than that of Fe_2O_3 phase. Besides, there was more coincidence in the white layer and Fe_3O_4 interface than in the white layer and Fe_2O_3 interface. Also oxidizing at 500 °C, led to higher surface microhardness because of thicker oxide layer and the formation of more Fe_3O_4 phase compared to oxidizing at 400 °C.

Alsaran et al. [29] (see Sect. 2.1.5) investigated the effects of post-plasma oxidizing on the mechanical properties of AISI 5140 steel.

The maximum surface hardness was obtained for plasma nitrided sample due to the formation of hard ε-Fe_3N iron nitride. The oxide layer formation lowered the surface hardness.

Post-oxidizing treatment had a beneficial effect on the wear behavior of nitrided sample. The lowest wear rate and friction coefficient belonged to the sample oxidized for 15 and 30 min, respectively. All of the data obtained for mechanical properties as well as the layers' thicknesses are presented in Table 2.19. The friction coefficient and wear rate for the sample post-oxidized for 60 min were higher compared with that of other oxidizing times. This was because of the failure of oxide layer by spallation (as a result of sputtering) despite the growing oxide layer with prolonging oxidation.

The improvement of wear resistance was also observed by SEM observation. As seen in Fig. 2.44, the wear track of the sample only nitrided was wider than that of nitrided-oxidized sample. Besides, some abrasive particles were also seen for just nitrided sample. This was attributed to the presence of a hard and brittle compound layer composed of two nitride phases.

Alsaran et al. [30] (see Sect. 2.1.5) studied the effects of post-aging treatment on mechanical properties of AISI 4140 steel.

Table 2.19 Properties of plasma nitrided and post-oxidized AISI 5140 steel [29]

Treatments	Compound layer (μm)	Oxide layer (μm)	Surface hardness ($HV_{0.025}$)	Surface roughness (R_a)	Friction coefficient	Wear rate (mm^3/ Nm × 10^6)
Nitrided	5.5–6.5	–	945–1000	0.15	0.38–0.41	4.84
Nitrided + 15 min post-oxidized	5–6	0.3–0.35	500–530	0.17	0.27–0.3	2.50
Nitrided + 30 min post-oxidized	5–5.5	0.35–0.4	480–520	0.16	0.23–0.25	3.22
Nitrided + 60 min post-oxidized	2–2.5	0.5–0.7	380–420	0.18	0.3–0.32	4.29

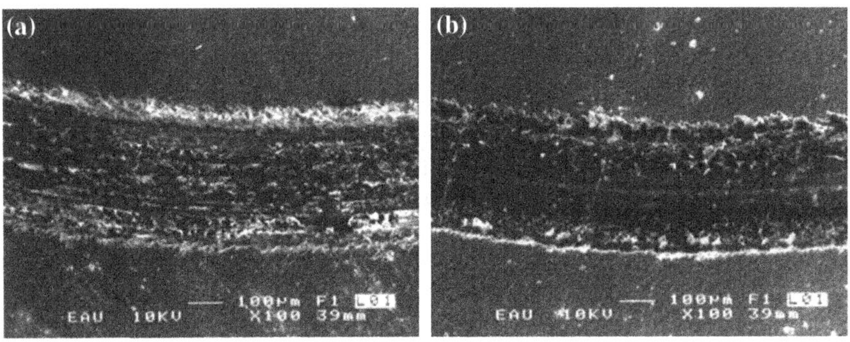

Fig. 2.44 SEM images of samples **a** plasma nitrided and **b** post-oxidized for 30 min after plasma nitriding [29]

The positive effect of post-aging treatment on the hardness was reported. The fact that hardness was high was related to the size and main crystal structure of α″. The post-aging treatment caused the lattice to be distorted and consequently the dislocation movement to be hindered and the hardness to be increased. The maximum hardness was achieved by duration of 1 h. The hardness was increased by post-aging duration until 1 h, however decreased after that due to the overaging. In this stage, the α″ phase was decomposed and therefore the coherency between the main lattice structure and this phase begun to disappear. The minimum hardness was achieved by duration of 12 h due to the growing of precipitates.

The minimum friction coefficient of 0.4 was achieved for post-aging duration of 0.5 h, while the maximum friction coefficient of 0.6 was obtained for a post-aging duration of 12 h. The minimum wear track width was observed for duration of 1 h. Because of higher hardness of compound layer formed on the surface of nitrided-quenched samples, the wear rate and abrasive particles were higher for this sample. The post-aging treatment resulted in the lower wear rate due to the hardening of diffusion layer. Also the lowest wear rate belonged to the sample post-aged for 1 h.

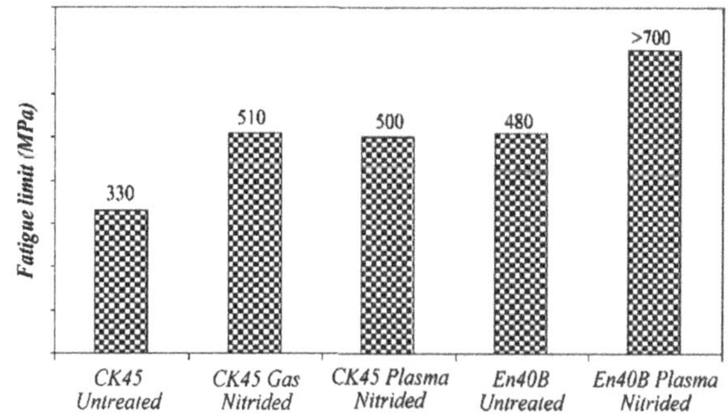

Fig. 2.45 Fatigue limits of the samples [32]

Ashrafizadeh [32] (see Sect. 2.1.5) compared the plasma and gas nitriding of Ck45 and En40B steels.

The lowest hardness of plasma nitrided Ck45 steel achieved was 470 HV and that of En40B was above 900 HV with a steep gradient. The hardness of compound layers of Ck45 and En40B was in the range of 500–800 HV and 900–1100 HV, respectively. This difference referred to the amount of absorbed nitrogen and type of nitrides formed in these two steels that were nitrides of iron and alloying elements in Ck45 and En40B, respectively. The lower nitriding depth of En40B was attributed to the higher content of carbon and alloying elements which had inhibiting effect on the nitrogen diffusion.

Fatigue limits of the untreated and nitrided samples are presented in Fig. 2.45. The improvement in fatigue limit for Ck45 steel was 52 % and for En40B was 46 % and higher depending on the treatment time. However for gas nitriding and Ck45 steel, no improvement of fatigue limit was observed even for duration of 70 h. Also it was found that compound layer had no significant effect on the fatigue limit in contrast to diffusion zone. The improvement of fatigue limit was due to two main reasons: higher case hardness and compressive residual stress. The increment of residual stress was caused by nitrogen being taken into solution in the matrix and the formation of nitride precipitates. It was concluded that gas nitriding is not suitable to improve the mechanical properties as it needs longer durations and produces thicker compound layers which is not suitable for some applications.

2.3 Corrosion Behavior

2.3.1 Austenitic Stainless Steels

Olzon-Dionysio et al. [7] (see Sect. 2.1.1) found that, the nitrided sample C (region 6 in Fig. 2.7) showed a more positive E_{corr} value than the nitrided sample B (regions 1 and 2). Moreover, the E_{corr} value of the unnitrided sample was lower than that of nitrided sample.

The corrosion current density value (for $E = 1.0$ V) obtained for sample C was about 8.0×10^3 μA cm^{-2} while it was about 9×10^4 μA cm^{-2} for sample B. This showed the lowering of the current density for this sample, indicating a lower pits dissolution rate. This may be caused by two factors

(1) The concentration of the hexagonal ε ($Fe_{2+x}N$) phase in the center (sample C, 38 %) was larger than in the border (sample B, 18.4 %). The hexagonal ε ($Fe_{2+x}N$) phase can reduce the dissolution rate of pits formed in the presence of aggressive ions.
(2) The nitrogen concentration in sample C was larger than in sample B, which changed pH and had a neutralizing effect on the acid pits of the corrosion surface.

Therefore, the sample with a ring showed lower corrosion resistance compared to the sample without it. Sample C showed the best corrosion resistance.

Egawa et al. [5] (see Sect. 2.1.1) found that for austenitic stainless steel, low temperatures (according to critical temperatures presented in Table 2.2) should be used to hamper the formation of dark phase discussed before. Also, the current density to maintain passivity on the plasma nitrided AISI 316 is higher than that of the untreated one. However in the case of other steels, plasma nitriding deteriorated corrosion resistance.

Nakajima et al. [8] (see Sect. 2.1.1) studied the effect of solution treatment after plasma nitriding on corrosion behavior of AISI 304.

After doing the anodic polarization test, pitting was reported as the main corrosion mechanism in the samples. The main reason for corrosion was found to be the formation of chromium nitride and consequently the depletion of boundaries. Also the surface of untreated sample was occupied uniformly by many corrosion pits.

2.3.2 *Martensitic Stainless Steels*

Li et al. [13] (see Sect. 2.1.2) observed that untreated AISI 410 sample did not show passivation layer and the current density increased with a high rate after the corrosion potential (Fig. 2.46). This could be due to the breakdown of the passive layer and localized corrosion. Also the untreated AISI 410 steel sample was severely experienced crevice corrosion and pitting corrosion during the polarization test (Fig. 2.47a).

As seen in Fig. 2.40, after plasma nitriding, the corrosion potential (E_{corr}) increased from −312 mV for the unnitrided sample to −190, −30, and −205 mV for the samples nitrided at 420, 460, and 500 °C nitrided samples, respectively.

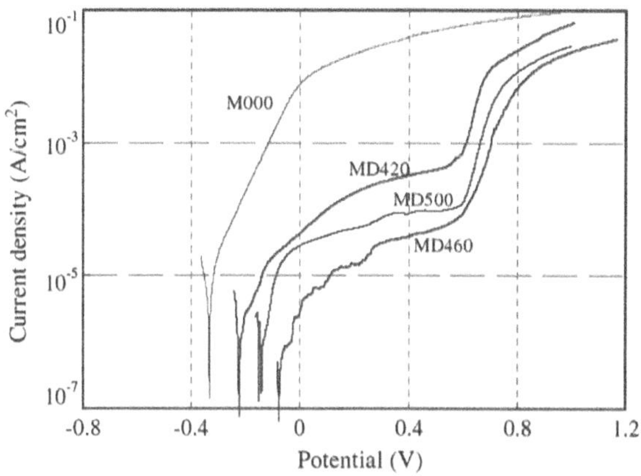

Fig. 2.46 Polarization curves of unnitrided and plasma nitrided AISI 410 steel in 3.5 wt% NaCl solution [13]

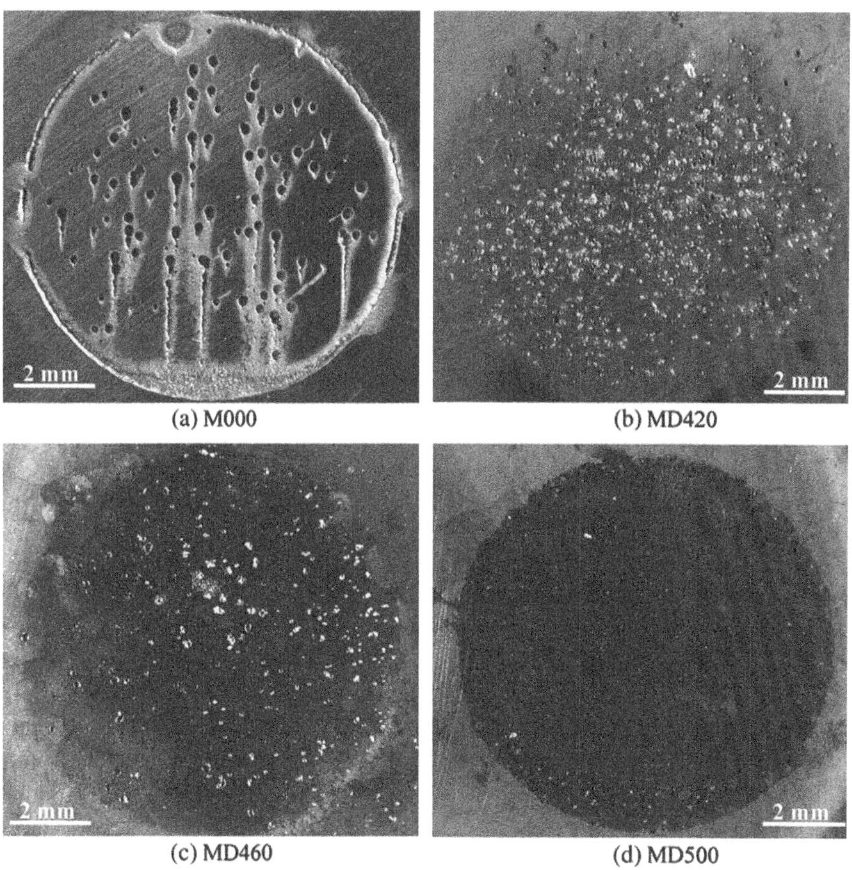

Fig. 2.47 Corrosion scars on **a** unnitrided and **b** 420 °C, **c** 460 °C, and **d** 500 °C nitrided AISI 410 martensitic stainless steel surface after electrochemical corrosion tests in 3.5 % NaCl solution [13]

It can be inferred from current densities in Fig. 2.46 that, corrosion rate for nitrided samples was much lower than that of the unnitrided one. It can also be indicated from this figure that pitting also occurred for all the nitrided samples which can be inferred from the sudden increase in current density after the passivation region. The pitting potential was about 600 mV for all the nitrided samples while it is much higher than −250 mV for the unnitrided sample.

It was reported that corrosion resistance of the AISI 410 martensitic stainless steel was improved by plasma nitriding indicated by the higher corrosion potential, higher pitting potentials, and lower current density of the nitrided samples. This improvement was also confirmed by corroded surface examination after polarization tests. Figure 2.47 shows that no crevice corrosion occurred in nitrided sample and the corrosion pits on a nitrided surface were smaller in size and fewer in number in comparison with the unnitrided sample. Besides, nitriding at 420 °C resulted in less resistance to pitting corrosion than the 460 and 500 °C. This completely agrees with Fig. 2.46 that the sample MD420 had the highest current densities among all nitrided samples.

The results of immersion tests in 1 % HCl solution also confirmed that plasma nitriding led to improving corrosion resistance of AISI 410 martensitic stainless steel. The average corrosion rate for unnitrided sample was 478 g m^{-2} d^{-1} in the 1 % HCl solution. It decreased to 141, 246, and 307 g m^{-2} d^{-1} for samples nitrided at 420, 460, and 500 °C, respectively. The iron nitrides formed on AISI 410 stainless steel surface could be responsible for good corrosion resistance and protect the substrate from being attacked by corrosive environment. The formation of an iron nitride compound layer was believed to be the main reason for improving corrosion resistance of the AISI 410 after plasma nitriding. With formation of a compound layer composed of γ' and/or ε iron nitride on the plasma nitrided surface, the effectiveness of chromium nitride formation in the corrosion resistance of the AISI 410 stainless steel faded.

In contrast to the results of electrochemical tests, the least corrosion rate belonged to the sample MD420 with the lowest weight loss among three nitrided samples. This inconsistence can rise from differing of test methods.

The surface of unnitrided sample was rough and severe pitting corrosion occurred after immersion test. However, the surface of nitrided samples was smooth.

Despite the chromium nitride precipitation at 460 and 500 °C, electrochemical and immersion corrosion tests revealed improving the corrosion resistance of the AISI 410 martensitic stainless steel at all three temperatures. This showed that the chromium nitride precipitation criteria for corrosion resistance of austenitic stainless steels may not be applicable to martensitic stainless steels.

For the untreated and 420 °C nitrided samples, the chromium content was higher in the near surface region probably due to the formation of chromium oxide passive film on the surface. The chromium content in the 460 °C treated surface was higher in the near surface region. However, precipitation of chromium nitride and the consequent depletion of free chromium in the iron matrix may have hindered the formation of a protective passive film. When nitriding at 500 °C, passive film was not formed on the nitrided surface because of (1) the lower chromium content in the top surface and (2) the chromium nitride precipitation.

2.3.3 Precipitation Hardening Stainless Steels

Sobiecki et al. [17] (see Sect. 2.1.3) found that the corrosion resistance of H9S2 steel was improved with plasma nitriding. This was inferred from the corrosion potential shifting from −250 to −500 mV. However, the corrosion current values were slightly increased compared to the untreated steel due to the formation of chromium nitride precipitations during nitriding above 500 °C.

Esfandiari et al. [16] (see Sect. 2.1.3) observed the improvement of corrosion resistance in the case of A286 precipitation hardening stainless steel by plasma nitriding (as shown in Fig. 2.48). However, it is clear from corrosion current densities that at temperatures below 420 °C, increasing the process time decreased corrosion resistance.

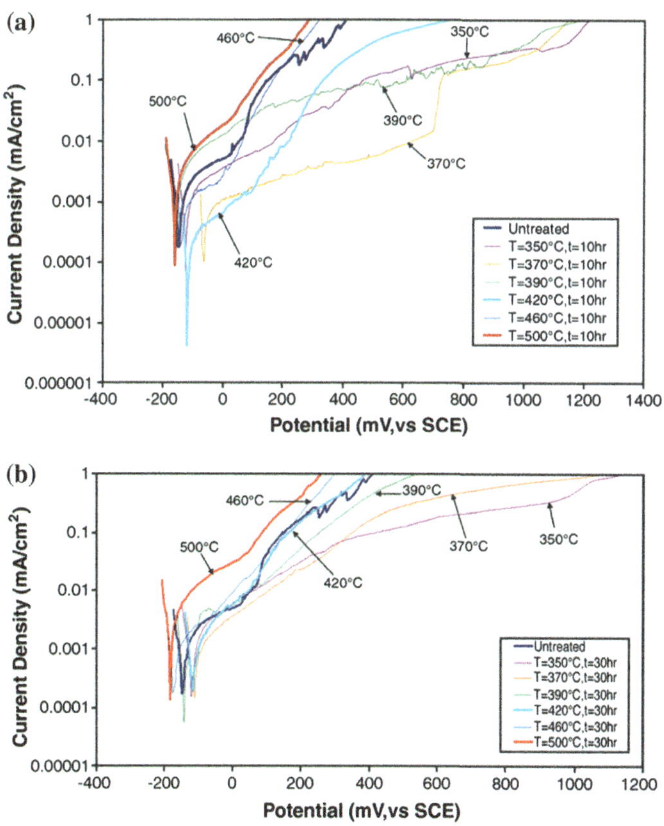

Fig. 2.48 Corrosion behavior of A286 samples plasma nitrided for **a** 10 h and **b** 30 h [16]

The evaluation of corrosion behavior of nitrided A286 precipitation hardening stainless steel in NaCl solutions showed two results: (1) nitriding at low temperature led to better pitting resistance than the untreated material and (2) nitriding at high temperature significantly deteriorated the corrosion resistance of A286 stainless steel. Also, results of corrosion test clearly indicated that plasma nitriding at low temperature, improved corrosion resistance of nitrided A286 stainless steel compared to the untreated sample because of the elimination of pitting corrosion in presence of a chloride-containing solution (Fig. 2.48). Thus, it was concluded that high nitrogen content corresponded for the above-mentioned improved corrosion resistance due to the formation of s-phase layer. However, it should be mentioned that, the beneficial effect of nitrogen only appeared when nitrogen is in solid solution, and the nitrided layer was a single phase. So that, CrN formation deteriorated the corrosion resistance of the stainless steel.

Corrosion-wear resistance of the untreated A286 steel was very low. This was proved by deep (18 μm) and wide wear tracks on the sample surface as seen in Fig. 2.49a, whereas, wear tracks on the surface of plasma nitrided samples were shallow (5–9 μm) and superficial as seen in Fig. 2.49b. Corrosion-wear resistance of 350 °C/10 h nitrided sample was improved only about 25 % related to the untreated one.

Although the 420 °C/10 h treated samples showed better corrosion resistance, the corrosion-wear resistance of the 420 °C/10 h and 500 °C/10 h treated samples were the same. Therefore, both layer hardness and thickness played an important role in the corrosion-wear process.

2.3.4 Tool Steels

Nouveau et al. [21] (see Sect. 2.1.4) evaluated corrosion properties of 90CrMoV8 tool steel used for wood machining. For testing corrosion properties the authors simulated the liquid oozed out during wood machining called "wood juice."

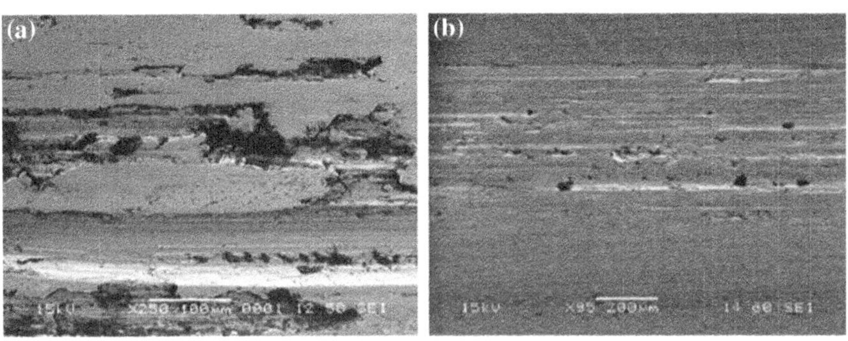

Fig. 2.49 SEM micrographs of the wear tracks on **a** untreated and **b** 420 °C/10 h treated A286 stainless steel after corrosion-wear test (10 N, 10 h) [16]

In electrochemical measurements, untreated steel behaved as a passive material which was characterized by a high corrosion potential value (−307 mV/ECS), a very low corrosion rate (~37 nA/cm^2) as well as a wide passive domain of about 250 mV. However, after plasma nitriding treatment, the extent of the passive domain for nitrided surfaces was reduced (~150 mV) indicating that the corrosion tendency was higher in terms of more negative potential and increased corrosion rates. The passivation of treated surfaces owed to the predominance of ε-phase nitrides and also to the absence of CrN precipitates. For too short nitriding times (6 h), surface experienced pitting, while prolonging the treatment (12 h), led to the cracking of the corroded surface. Therefore, it showed that long duration treatments were recommended for achieving a thick very hard layer, which is interesting from a tribological point of view. Furthermore, due to the release of potentially aggressive liquids during the wood machining, tools need also to be corrosion resistant (Fig. 2.50 and Table 2.20). Therefore, the best nitriding condition was 500 °C for 10 h.

2.3.5 Structural and Constructional Steels

Yasavol et al. [28] (see Sect. 2.1.5) evaluated the post-oxidizing treatment effects on corrosion properties of AISI 4130 steel seen in Table 2.21. Post-oxidizing treatment using 100 % O_2 led to lower corrosion resistance compared to other atmospheres despite the thickest oxide layer formed using this atmosphere. This behavior was attributed to the presence of both Fe_2O_3 and Fe_3O_4. However, the sample oxidized under atmosphere of 25 % O_2 + 75 % H_2 showed a better corrosion resistance due to the formation of a dense oxide layer despite its lower thickness compared to 100 % O_2. The decrement of oxygen content led to even lower corrosion rate. However, an increase in the oxidizing temperature to 500 °C degraded the corrosion resistance due to the formation of a thin and non-compact oxide layer.

In all of the oxidizing atmospheres, the corrosion resistance was improved compared to only plasma nitriding. This was due to the formation of a dense oxide layer and filling the pits of compound layer.

Alsaran et al. [29] (see Sect. 2.1.5) also studied the effects of post-oxidizing treatment on corrosion behavior of AISI 5140 steel.

It was observed that both plasma nitriding and plasma nitriding-post-oxidizing treatments significantly improved the corrosion resistance of the samples. The untreated samples showed a low corrosion potential (−720 mV) and very high anodic currents indicating an active behavior. For all of the treated samples a good resistance to general corrosion was observed in comparison with untreated specimen, however all of them were subjected to pitting corrosion. As seen in Fig. 2.51, after the potentiodynamic polarization tests, corrosion pits resulting from local passivity breakdown were distributed on the surface. This behavior could be ascribed to the presence of hematite as the main constituent of oxide layer due to the

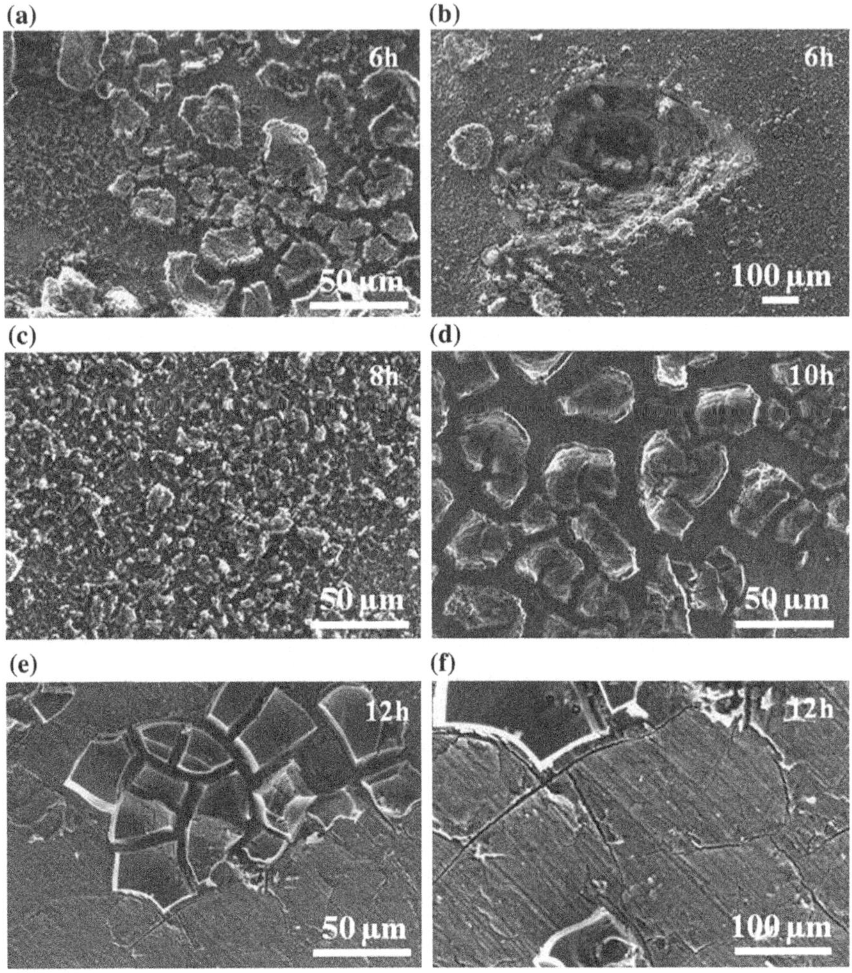

Fig. 2.50 SEM micrographs showing the surface morphologies of corroded nitrided steel samples after 24 h corrosion tests [21]

Table 2.20 Electrochemical parameters of the different samples deduced from polarization curves [21]

	Untreated steel	6 h at 500 °C	8 h at 500 °C	10 h at 500 °C	12 h at 500 °C
E_{corr} (mV/ECS)	−307	−559	−569	−582	−575
I_{corr} (nA/cm^2)	37	233	417	392	347
Passive domain extent (mV)	255	87	150	157	156

Table 2.21 The results of corrosion test on nitrided-post-oxidized samples [28]

Plasma oxidizing treatment parameters	I_{corr} (A/cm^2)	E_{corr} (V)	Corrosion rate (mm/year)
500 °C, 25 % O_2 + 75 % H_2	6.92×10^{-9}	−0.09	0.0057
500 °C, 10 % O_2 + 90 % H_2	3.39×10^{-8}	−0.20	0.0121
500 °C, 100 % O_2	8.87×10^{-8}	−0.30	0.0089
400 °C, 25 % O_2 + 75 % H_2	9.09×10^{-7}	−0.39	0.0077

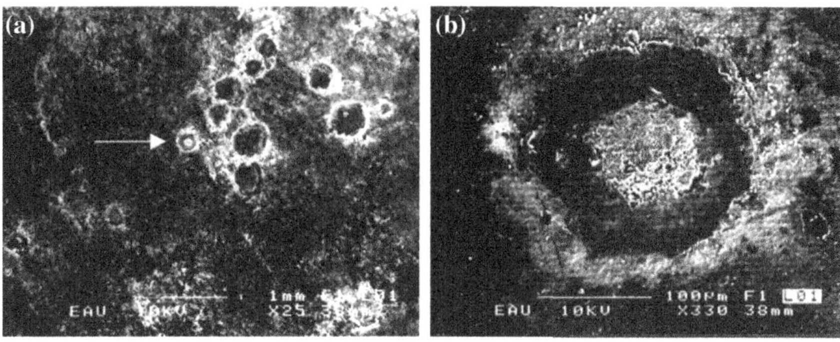

Fig. 2.51 SEM micrographs showing the surface of corroded sample post-oxidized for 60 min **a** after plasma nitriding and **b** one pit [29]

porous and low-adherent nature of hematite which was grown during oxidation treatment. The higher corrosion potential (−255 mV) belonged to the sample post-oxidized for 15 min which was higher than that of the plasma nitrided sample. The lowest anodic currents were obtained for durations of 15 and 30 min whereas the highest anodic currents were observed for duration of 60 min. This was owing to the fact that increasing of oxidation time led to the spallation and consequently failure of the oxides.

Alsaran et al. [30] (see Sect. 2.1.5) studied the effects of post-aging treatment on corrosion resistance of plasma nitrided AISI 4140 steel.

Among post-aged samples, the best result was obtained for post-aging duration of 1 h. The corrosion resistance was inversely related with duration. However the corrosion resistance of just plasma nitrided sample was higher than quenched and post-aged samples.

References

1. Xu X L, Wang L, Yu Z W, Hei Z K (2000) Microstructural characterization of plasma nitrided austenitic stainless steel. Surf Coat Technol 132:270-274
2. Wang J, Xiong J, Peng Q, Fan H, Wang Y, Li G, Shen B (2009) Effects of DC plasma nitriding parameters on microstructure and properties of 304L stainless steel. Mater Charact, 60:197–203

3. Stinville J C, Villechaise P, Templier C, Riviere J P, Drouet M (2010) Plasma nitriding of 316L austenitic stainless steel: Experimental investigation of fatigue life and surface evolution. Surf Coat Technol 204:1947–1951
4. Li G J, Peng Q, Li C, Wang Y, Gao J, Chen S Y, Wang J, Shen B L (2008) Effect of DC plasma nitriding temperature on microstructure and dry-sliding wear properties of 316L stainless steel. Surf Coat Technol 202:2749–2754
5. Egawa M, Ueda N, Nakata K, Tsujikawa M, Tanaka M (2010) Effect of additive alloying element on plasma nitriding and carburizing behavior for austenitic stainless steels. Surf Coat Technol 205:S246–S251
6. de Oliveira A M, Muñoz Riofano R M, Casteletti L C, Tremiliosi G F, Bento C A S (2003) Effect of the Temperature of Plasma Nitriding in AISI 316L Austenitic Stainless Steel. Revista Brasileira de Aplicações de Vácuo 22:63-66
7. Olzon-Dionysio M, Campos M, Kapp M, de Souza S, de Souza S D (2010) Influences of plasma nitriding edge effect on properties of 316 L stainless steel. Surf Coat Technol 204:3623–3628
8. Nakajima M, Nakamura Y, Suzuki K, Bai Y, Uematsu Y (2014) Effect of solution treatment after nitriding on fatigue properties in type 304 stainless steel. Int J Fatigue 68:103–110
9. Wang L, Ji S, Sun J (2006) Effect of nitriding time on the nitrided layer of AISI 304 austenitic stainless steel. Surf Coat Technol 200:5067 – 5070
10. Sun Y, Haruman E (2006) Effect of carbon addition on low-temperature plasma nitriding characteristics of austenitic stainless steel. Vac 81:114–119
11. Borges C F M, Hennecke S, Pfender E (2000) Decreasing chromium precipitation in AISI 304 stainless steel during the plasma-nitriding process. Surf Coat Technol 123:112–121
12. Alphonsa I, Chainani A, Raole P M, Ganguli B, John P I (2002) A study of martensitic stainless steel AISI 420 modified using plasma nitriding. Surf Coat Technol 150:263–268
13. Li C X, Bell T (2006) Corrosion properties of plasma nitrided AISI 410 martensitic stainless steel in 3.5% NaCl and 1% HCl aqueous solutions. Corros Sci 48:2036–2049
14. Pinedo C E, Monteiro W A (2004) On the kinetics of plasma nitriding a martensitic stainless steel type AISI 420. Surf Coat Technol 179:119–123
15. Wu K, Liu G Q, Wang L, Xu B F (2010) Research on new rapid and deep plasma nitriding techniques of AISI 420 martensitic stainless steel. Vac 84:870–875
16. Esfandiari M, Dong H (2007) Improving the surface properties of A286 precipitation-hardening stainless steel by low-temperature plasma nitriding. Surf Coat Technol 201:6189–6196
17. Sobiecki J R, Ma!nkowski P, Patejuk A (2004) Improving the performance properties of valve martensitic steel by glow discharge-assisted nitriding. Vac 76:57–61
18. Li G J, Wang J, Li C, Peng Q, Gao J, Shen B (2008) Microstructure and dry-sliding wear properties of DC plasma nitrided 17-4 PH stainless steel. Nucl Instrum Methods Phys Res B 266:1964–1970
19. Dong H, Esfandiari M, Li X Y (2008) On the microstructure and phase identification of plasma nitrided 17-4PH precipitation hardening stainless steel. Surf Coat Technol 202:2969–2975
20. Birol Y (2010) Response to thermal cycling of plasma nitrided hot work tool steel at elevated temperatures. Surf Coat Technol 205:597–602
21. Nouveau C, Steyer Ph, Rao K R M, Lagadrillere D (2011) Plasma nitriding of 90CrMoV8 tool steel for the enhancement of hardness and corrosion resistance. Surf Coat Technol 205: 4514–4520
22. da Silva Rocha A, Strohaecker T, Tomala V, Hirsch T (1999) Microstructure and residual stresses of a plasma-nitrided M2 tool steel. Surf Coat Technol. 115:24–31
23. Pessin M A, Tier M D, Strohaecker T R, Bloyce A, Sun Y, Bell T (2000) The effects of plasma nitriding process parameters on the wear characteristics of AISI M2 tool steel. Tribol Lett 8:223–228
24. Li C X, Sun Y, Bell T (2000) Factors influencing fretting fatigue properties of plasma-nitrided low alloy steel. Mater Sci Eng A292:18–25

25. Karaog˘lu S (2003) Structural characterization and wear behavior of plasma-nitrided AISI 5140 low-alloy steel. Mater Charact 49:349– 357
26. Karakan M, Alsaran A, Çelik A (2003) Effects of various gas mixtures on plasma nitriding behavior of AISI 5140 steel. Mater Charact 49:241– 246
27. Mahboubi F, Abdolvahabi K (2006) The effect of temperature on plasma nitriding behaviour of DIN 1.6959 low alloy steel. Vac 81:239–243
28. Yasavol N, Mahboubi F (2012) The effect of duplex plasma nitriding-oxidizing treatment on the corrosion resistance of AISI 4130 steel. Mater Des 38:59–63
29. Alsaran A, Altun H, Karakan M, Çelik A (2004) Effect of post-oxidizing on tribological and corrosion behaviour of plasma nitrided AISI 5140 steel. Surf Coat Technol 176:344–348
30. Alsaran A, Yildiz F, Çelik A (2006) Effects of post-aging on wear and corrosion properties of nitrided AISI 4140 steel. Surf Coat Technol 201:3147–3154
31. Alves Jr C, Lima J A, Hajek V, Marimon da Cunha J B, dos Santos C A (2007) Effect of cooling rate on properties of plasma nitrided AISI 1010 steel. Surf Coat Technol 201:7566–7573
32. Ashrafizadeh F (2003) Influence of plasma and gas nitriding on fatigue resistance of plain carbon (Ck45) steel. Surf Coat Technol 173 –174:1196–1200
33. Marchev K, Cooper C V, Giessen B C (1998) Observation of a compound layer with very low friction coefficient in ion-nitrided martensitic 410 stainless steel. Surf Coat Technol 99:229-233
34. Şengül A B, Çelik A (2011) Effect of plasma nitriding on fatigue crack growth on AISI 4140 steel under variable amplitude loading. Surf Coat Technol 205:5172–5177

Chapter 3
Pulsed DC Glow Discharge Plasma Nitriding

Pulsed power supply is used for plasma nitriding to overcome the problems of conventional DC plasma nitriding. Therefore, using this power source has provided some noticeable advantages such as the following:

1. A more accurate control on nitriding process
2. Controlling and lowering the process temperature simply by adjusting the pulse width without any change in bias voltage
3. Avoiding the arcing phenomenon on the surface of the workpiece
4. A more uniform temperature distribution and lowering the overheating risk of the surface of the workpiece.

Accordingly, many studies have been done to exactly investigate the role of pulsed DC current and its parameters (specially pulse width) in pulsed DC plasma nitriding to optimize them and have the best surface properties [1–4].

Here are presented the results of some studies which have been done about the effects of pulsed DC plasma nitriding on microstructural, mechanical, and corrosion properties of steels.

3.1 Microstructure

3.1.1 Austenitic Stainless Steels

Liang [5] plasma nitrided AISI 304 austenitic stainless steel with the following parameters: temperature of 350–540 °C, atmosphere of NH_3, pressure of 300 Pa, and duration of 4 h. For all the samples, the white nitrided layer was observed. It was observed that, increasing temperature led to thicker nitrided layer.

H. Aghajani and S. Behrangi, *Plasma Nitriding of Steels*, Topics in Mining, Metallurgy and Materials Engineering, DOI 10.1007/978-3-319-43068-3_3

For samples nitrided at 430–450 °C, a surface nitride layer of around 7–12 μm thickness was clearly seen while for 350 °C, the nitrided layer was 2–3 μm. In the case of nitriding at 465 °C, a 20 μm thick nitrided layer with some chromium nitride precipitates was obtained.

For samples treated at lower temperatures (below 400 °C), γ_N and γ phases were detected. The sample treated below 400 °C exhibited a series of broad XRD peaks, shifted to lower 2θ values related to expanded austenite (as seen in conventional DC plasma nitrided austenitic stainless steel in Chap. 2) that is a major constituent in the nitrided layers. The shift degree of expanded austenite increased with processing temperature increase up to 420 °C. For the higher nitriding temperature about 460 °C, chromium nitride and iron nitrides appeared in the nitrided layer.

Singh et al. [6] investigated the effect of surface roughness on the plasma nitriding behavior of the AISI 304 austenitic stainless steel. They prepared the samples in four different surface finishes: (a) polished samples (R_a = 0.05), (b) rough polished samples (R_a = 0.075), (c) machined samples (R_a = 0.47), (d) sample after grinding (R_a = 1.02). They plasma nitrided samples using these parameters: temperature of 560 °C, duration of 24 h, voltages of 540 and 620 V, atmospheres of 20 % N_2 + 80 % H_2 and vice versa, and pressure of 400 Pa.

According to Fig. 3.1, with changing nitriding atmospheres, different thicknesses for nitrided layer were obtained. Under the atmosphere of 80 % N_2, a case depth of 110 μm and 50 μm was observed in mirror polished and ground sample, respectively. However, for 20 % N_2, they were 90 and 40 μm, respectively.

After plasma nitriding, ε-Fe_3N, γ′-Fe_4N, CrN, γ and α phases were observed for all the four samples. In the case of mirror polished samples, the value of Fe_3N and Fe_4N was higher when treated with 80 % nitrogen. For ground samples the value of α phase was higher.

α (ferrite) was present on the top surface of all the unnitrided samples. This was probably due to the stress induced γ to α transition during machining and polishing. However, pure γ phase was observed for ground sample means that stress-induced

Fig. 3.1 Optical microscope images of **a** mirror polished sample plasma nitrided at 80 % N_2 + 20 % H_2, **b** mirror polished sample plasma nitrided at 20 % N_2 + 80 % H_2, **c** ground sample plasma nitrided at 80 % N_2 + 20 % H_2, and **d** ground sample plasma nitrided at 20 % N_2 + 80 % H_2 [6]

layer was taken off. After plasma nitriding, α had the most amount for ground sample in both gas composition.

On the other hand, the relative intensity of nitride phases (Fe_3N, Fe_4N, CrN) peaks with respect to γ phase peak was the least in ground sample indicating that nitride formation was very less, and was maximum on mirror polished sample. From the above results, it can be concluded that pure γ phase hinders nitride formation while α phase on the top surface plays a beneficial role to facilitate nitride formation.

The less formation of iron nitrides on the surface of ground sample could be resulted by the less nitrogen diffusion. This can be ascribed to the higher packing fraction of austenite in comparison with ferrite.

Sharma et al. [7] compared plasma nitriding behavior of AISI 302 and 304 austenitic stainless steels in pure N_2 and N_2–H_2 atmosphere according to Table 3.1.

In this study, it was observed that, pure nitrogen cannot act as a good plasma nitriding atmosphere. Because in the absence of hydrogen, iron nitrides such as Fe_3N and Fe_4N were not formed indicating the key role of hydrogen in this process. This effect can be ascribed to the removal of surface oxide from the surface and consequently increasing nitrogen atoms diffusion into the surface and sputtering of iron atoms over the surface.

The amount of γ-Fe in AISI 304 sample plasma nitrided in N_2–H_2 atmosphere, was relatively higher than α-Fe, whereas using pure nitrogen as plasma nitriding atmosphere resulted in relatively lower amount of γ-Fe than α-Fe. It was observed that with increasing the nitriding frequency from 50 to 200 kHz, the amount of CrN, Fe_2O_3, and α-Fe phases gradually decreased.

Gontijo et al. [8] also observed similar phases reported by other authors. They plasma nitrided AISI 304L samples at 350, 400, 450, and 500 °C, in atmosphere of 20 % N_2 + 80 % H_2, under pressure of 500 Pa for 3 h.

It can be seen from Fig. 3.2 that at lower temperatures (350 and 400 °C), a thin layer of s-phase can be observed on top of the substrate. A dark layer is also identified between these two. This layer was formed with diffusion of carbon to the interface between the transformed layer and the substrate. The thickness of s-phase layer increased with nitriding temperature.

For samples nitrided at lower temperatures (350 and 400 °C), the substrate austenite phase and expanded austenite phase were observed. For higher temperatures (450 and 500 °C), the CrN phase was also detected as well as expanded austenite. However, for thicker nitrided layers, substrate austenitic phase was not detected.

Table 3.1 Parameters used by Sharma et al. [7]

Discharge frequency (kHz)	Discharge voltage (V)	Discharge current (A)	Sample temperature (°C)	Nitriding duration (h)
50	636–644	1.19–1.27	540	4
100	631–636	1.20–1.27		
200	618–622	1.30		

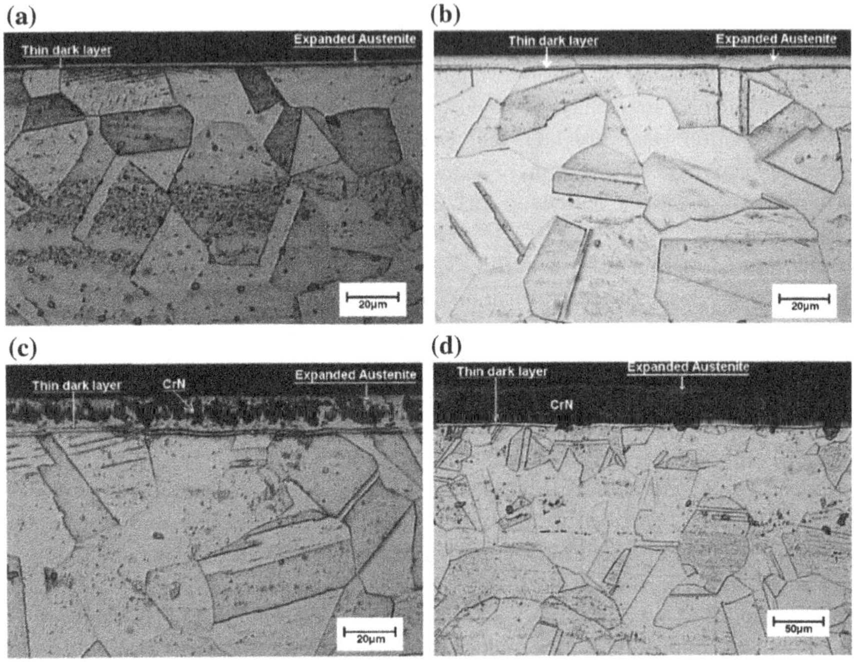

Fig. 3.2 Optical microscope cross-sectional images of the samples nitrided at **a** 350 °C, **b** 400 °C, **c** 450 °C, and **d** 500 °C [8]

The presence of s-phase and CrN in the microstructure of plasma nitrided AISI 304L austenitic stainless steel was also reported by Menthe et al. [9]. They have applied different temperatures in 375–475 °C range with different durations.

The decrement of nitrided layer with decreasing the nitrogen partial pressure of the treatment gas, the treatment duration and the treatment temperature was reported in this study.

Jeong et al. [10] observed the formation of s-phase at lower temperatures and CrN at higher temperatures after plasma nitriding of AISI 316L, as reported in the case of AISI 304 and 304L above. They applied these conditions: temperatures of 400 and 500 °C, duration of 5 h, atmosphere of 80 % N_2 + 20 % H_2, pressure of 533 Pa, voltage of 400 V and also different frequencies and pulse on/off times.

At 400 °C, just the main peaks of the s-phase were detected. At 500 °C, the expanded austenite was disappeared and the constituents of the nitride layer were Fe_4N and CrN with no sign of s-phase (Fig. 3.3). This transformation was the result of partial decomposition of s-phase. However, no changes in nitrides composition were observed with pulse frequency variations (50–5000 Hz).

According to Fig. 3.4, the effect of temperature on the compound layer thickness was significant but this effect was slight for pulse frequency (Both of them have direct relationship).

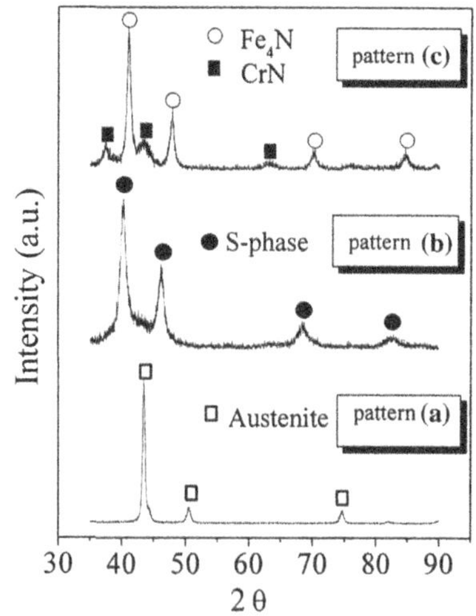

Fig. 3.3 XRD patterns of unnitrided and plasma nitrided AISI 316L austenitic stainless steel [10]

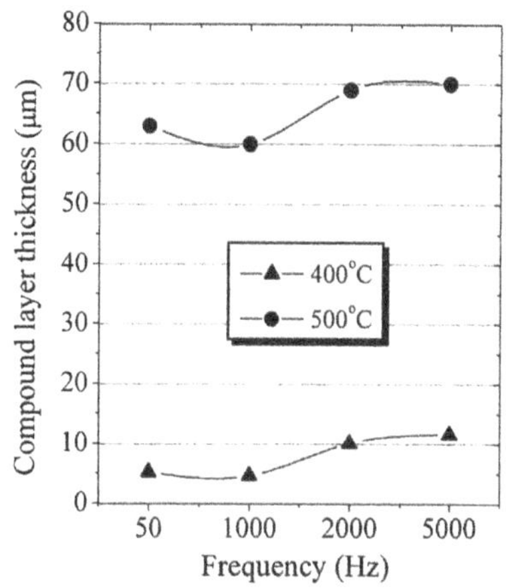

Fig. 3.4 Changes of the nitride layers thickness with frequency [10]

At 500 °C, a lot of microcracks were formed in the nitride layer that was due to the internal stresses.

Gil et al. [11] also detected s-phase on the plasma nitrided surface of AISI 316L austenitic stainless steel. The applied parameters were: temperature of 410 °C, duration of 8 h, atmosphere of 25 % N_2 + 75 % H_2, pressure of 6.5 hPa, and voltage of 680 V.

Li et al. [12] observed that the so-called edge effect present in the convectional plasma nitrided samples (see Sect. 2.1.1 in Chap. 2) completely avoided with applying pulsed DC current.

They also observed that the nitrided layer formed at low temperatures (below 480 °C) mainly constituted by expanded austenite phase. While plasma nitriding at high temperature (above 510 °C) led to the formation of a 3-layered structure consisting of 1) a top compound (γ'-Fe_4N + γ_N + CrN) layer, 2) an intermediate (γ'-Fe_4N + γ_N + CrN + α) layer, and 3) a lower layer.

Xia et al. [13] reported the existence of CrN, $Fe_{4-}N$, and Fe_3N at high temperature plasma nitrided 1Cr18Ni9Ti austenitic stainless steel. They treated samples at temperature of 560 °C, for 2 h and in atmosphere of 20 % N_2 + 80 % H_2.

It was reported that, the microstructure of pulsed DC plasma nitrided austenitic stainless steel was almost like that of conventional plasma nitrided austenitic stainless steel.

3.1.2 Martensitic Stainless Steel

Corengia et al. [14] studied microstructure of pulsed DC plasma nitrided AISI 410 martensitic stainless steel with these parameters: temperatures of 350, 400 and 500 °C, durations of 20 and 28 h, atmosphere of 25 % N_2 + 75 % H_2, pressure of 6 hPa and voltage of 750 V. There were two groups of samples according to kind of heat treatment before plasma nitriding: annealed samples and quenched-tempered ones.

In the case of quenched-tempered sample nitrided for 20 h at 400 °C, a white nitrided layer of approximately 30 μm thick was observed.

The microstructure of 350 °C annealed plasma nitrided sample consisted of mainly α-Fe and traces of γ'-Fe_4N. For the samples plasma nitrided for 20 h at 400 and 500 °C, it was observed that in both cases, the main components of the nitrided layer were γ'-Fe_4N, ε-$Fe_{2\text{-}3}N$, CrN and α-Fe. At high temperatures, the decomposition of expanded martensite into α and CrN was reported.

This also was reported by Corengia et al. [15].

Yun-tao et al. [16] observed the microstructure of plasma nitrided AISI 420 martensitic stainless steel with these treatment conditions: temperatures of 350 and 550 °C, duration of 15 h, atmosphere of 25 % N_2 + 75 % H_2, pressure of 600 Pa and voltage of 600 V.

The microstructure of treated samples was consisted of the top compound layer and the nitrogen diffusion beneath. The more the nitriding temperature, the thicker was the nitrided layer. The thickness of the compound layer at 350 and 550 °C were 90 and 130 μm.

It was reported that the surface of the sample nitrided at 350 °C consisted of mainly ε-Fe_3N with traces of α_N phase.

The surface of the sample nitrided at 550 °C was mainly composed of γ'-Fe_4N. The amount of ε-Fe_3N was much less than that in the surface of the sample nitrided at 350 °C. The α_N was not detected but the CrN was detected. This phase composition showed the decomposition of expanded phase into ferrite and CrN. It was reported that the lattice expansion was diminished and the XRD peak was shifted back to its position.

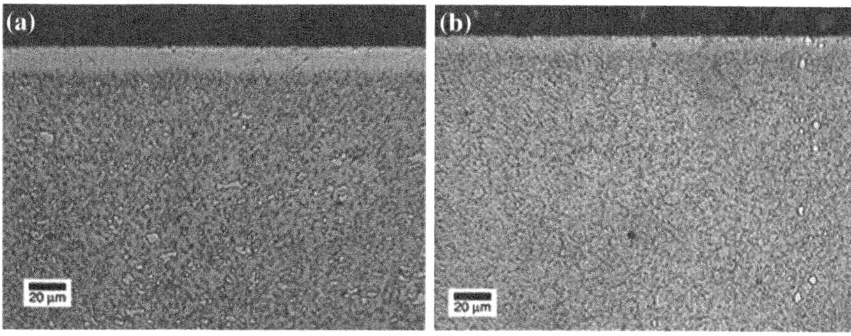

Fig. 3.5 Optical microscope images of nitrided layers of **a** N695 and **b** M340 [18]

Xi et al. [17] plasma nitrided AISI 420 martensitic stainless steel with these parameters: temperatures of 350, 450 and 550 °C, duration of 15 h, atmosphere of 25 % N_2 + 75 % H_2, pressure of 600 Pa and voltage of 600 V. Like other researchers they identified two well-defined layers on the surface of the nitrided samples: compound layer and diffusion zone. The average thicknesses of the compound layers were about 90 µm, 105 µm and 130 µm for the 350, 450, and 550 °C nitrided samples respectively. However, due to the absence of a clear boundary between two layers, the depth of diffusion layer was not reported.

The present phases in the microstructure of plasma nitrided samples was exactly like that of reported by Yun-tao et al. [16]. With increasing temperature, the amount of ε and $\alpha_{\text{-N}}$ decreased, however that of γ' increased and also CrN began to appear.

Brühl et al. [18] plasma nitrided M340 and N695 martensitic stainless steels with following parameters and observed their microstructure: temperature of 360 °C, duration of 10 h, atmosphere of 22 % N_2 + 78 % $H_{2,}$ and voltage of 450 V.

Some carbides with white and rounded structures were observed in the microstructure of these two steels (Fig. 3.5). Due to the formation of no dark sublayers above the white layer, it was concluded that Fe_4N and CrN were not precipitated. The white layer in nitrided martensitic steels is a stressed structure called "expanded martensite." However, in the authors' opinion this nomenclature is not very suitable since martensite is a stressed structure itself and therefore suggested "expanded ferrite" instead.

Chromium carbides were detected even before nitriding treatment. After nitriding, the α-Fe lattice was expanded and also chromium nitrides and γ nitrides were detected.

The plasma nitriding parameters used by Pinedo [19] (520 °C, 12 h, 50 % N_2 + 50 % H_2 and 200 Pa) resulted in the formation of no compound layer, therefore, the diffusion zone depth was considered as nitriding depth. Figure 3.6 shows the microstructure of 12 h plasma nitrided steel. As can be seen, a sharp and clear interface between the diffusion zone and the substrate is observable. The nitriding depths resulted by treatment durations of 12 and 8 h treatment were 125 and 105 µm, respectively.

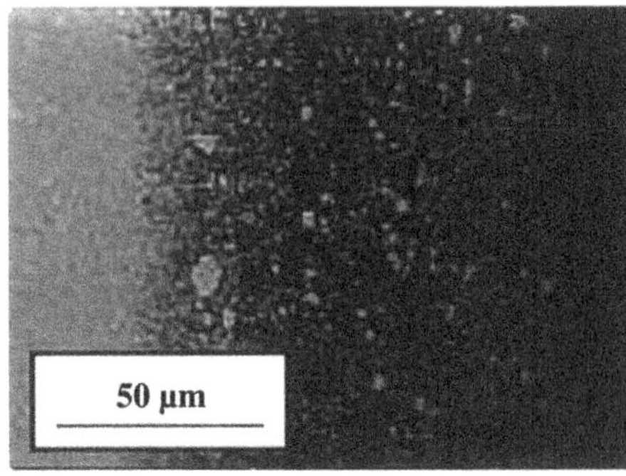

Fig. 3.6 Optical microscope image of the surface after plasma nitriding [19]

Espitia et al. [20] studied the plasma nitriding of AISI 410 and compared with ASTM A 743 grade CA6NM martensitic stainless steels. Nitriding treatments were done at 400 °C in atmosphere of 75 % N_2 + 25 % H_2 for 20 h.

The nitrided layer of AISI 410 had the thickness of about 25 μm. The first 5 μm under the surface was constituted by nitrides and expanded martensite (α_N). The nitrides were characterized as mainly ε and a few γ′. Beneath this first region, only expanded martensite was observed to the depth of about 25 μm.

3.1.3 *Precipitation Hardening Stainless Steels*

Brühl et al. [18] studied the microstructure of Corrax precipitation hardening stainless steel which was nitrided with following parameters: temperature of 360 °C, duration of 10 h, atmosphere of 22 % N_2 + 78 % H_2, and voltage of 450 V.

As seen in Fig. 3.7, no iron nitrides or carbides were formed after nitriding the Corrax samples but lattice expansion caused the Fe α peaks to shift to lower angles.

This expansion was observed in the case of austenitic and martensitic stainless steels.

3.1.4 *Ferritic Stainless Steels*

Alves Jr et al. [3] investigated the effect of varying t_{off} (the time between pulses) on the microstructure of pulsed DC plasma nitrided AISI 409 ferritic stainless steel with these parameters: temperature of 500 °C, atmosphere of 20 % N_2 + 80 % H_2, pressure of 600 Pa, and parameters listed in Table 3.2.

It was reported that from microstructural point of view, overall aspect of all samples were similar, regardless of t_{off}. The presence of both transformed and non-transformed regions was also reported.

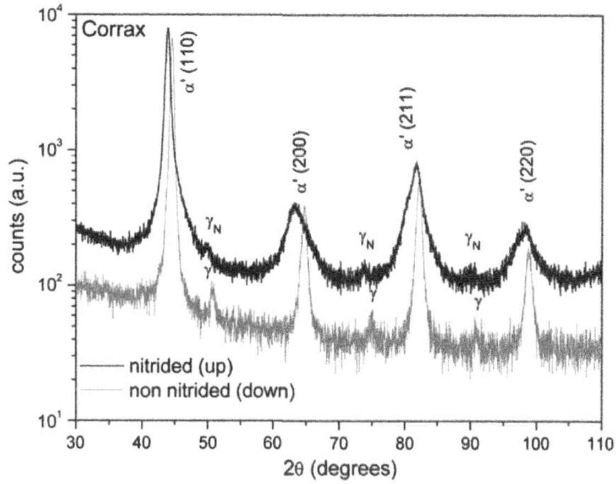

Fig. 3.7 Glancing angle WRD of Corrax before and after nitriding [18]

Table 3.2 Values of t_{off} investigated with corresponding voltages and current [3]

t_{off} (ms)	Voltage (V)	Current (mA)
0	500	493
1	515	500
2	527	485
3	523	480
4	513	499
5	514	492

The existence of both γ'-Fe_4N and ε-$Fe_{2-3}N$ phases was indicated by XRD patterns on the nitrided surfaces.

According to Fig. 3.8, in the case of $t_{off} = 0$ and 4 ms, the relative intensity of γ' and ε phases was higher than that of the α-Fe phase.

According to Fick's second law, the penetration depth of nitrogen depends on the diffusion coefficient of the matrix and on the concentration gradient of atomic nitrogen. The diffusion coefficient can be varied by the population of defects which is dependent on the rate of surface ionic sputtering. The concentration gradient can be adjusted by varying the concentration of atomic nitrogen at the surface of the specimen. In this case, nitrogen is produced either by the dissociation of N_2 molecules or by the transformation of the primary nitrides into secondary nitrides. As t_{off} increased, the sputtering rate decreased which led to the reduction in the number of defects and consequently, the diffusion coefficient. The concentration of atomic nitrogen resulting from the transformation of primary nitrides into secondary nitrides also changed, as there was enough time for the sequence $FeN \rightarrow Fe_{2-3}N \rightarrow Fe_4N$ to take place, with no interruption of the ionic sputtering process.

It was reported that, except for samples with $t_{off} = 1$ and 2 ms, the thickness of nitrided later was constant (about 60 μm).

It was observed that, the highest and the lowest nitrogen concentration in the surface, belonged to the sample with $t_{off} = 0$ ms (nitrided in steady state DC) and

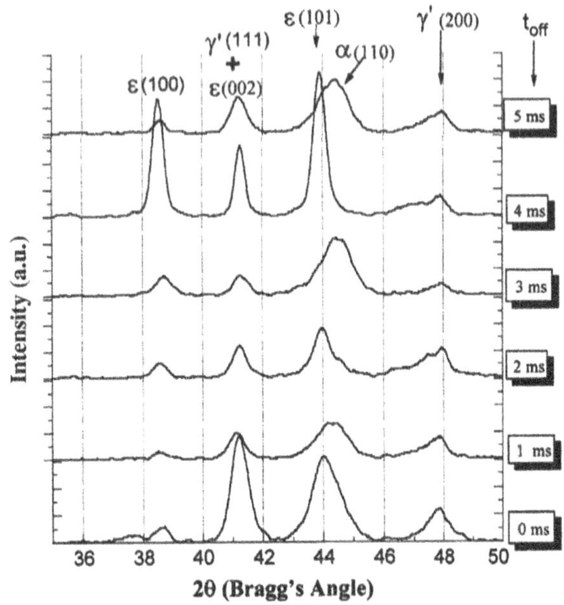

Fig. 3.8 XRD patterns of nitrided surfaces using different t_{off} values [3]

the sample with $t_{off} = 3$ ms, respectively. Therefore, it was concluded that the penetration depth of nitrogen was affected by t_{off}, specially at 0 and 2 ms, where the largest variations in the thickness of the nitrided layer were observed.

Larisch et al. [21] compared microstructural properties of X6Cr17 ferritic stainless steel, X2CrNiMoN22.5.3 austenitic-ferritic stainless steel, X8CrNiTi18.10 austenitic stainless steel and X5CrNi18.10 austenitic stainless steel after plasma nitriding with these parameters: temperature of 250–500 °C, duration of 8–60 h, pressure of 300 Pa, and voltage of 400–450 V. The atmosphere used consisted of 40–80 % N_2 + 10–30 % H_2 + 10–30 % Ar.

They found that the temperature, the nitriding time, and the steel grade are the main factors affecting layer thickness and nitrogen concentration. The higher nitrogen concentration values and larger diffusion depths were obtained at higher temperatures (Fig. 3.9) and longer nitriding times.

It was found that, the lowest nitrogen diffusion depth belonged to duplex (austenitic-ferritic) steel. Also, the nitrogen diffusion depth in the ferritic steel was higher than austenitic one.

Among different gas compositions, 80 % N_2 + 10 % H_2 + 10 % Ar had the lowest diffusion depth as can be seen in Table 3.3. Such a low diffusion depth was probably due to the lower proportions of hydrogen and argon gases. Hydrogen can improve the energy of ions to sputter and remove the surface sufficiently. Also, the addition of argon to the nitriding atmosphere resulted in the higher current density and consequently higher plasma density. Also, higher voltage leading to stronger bombardment of the surfaces improved the results (Table 3.3).

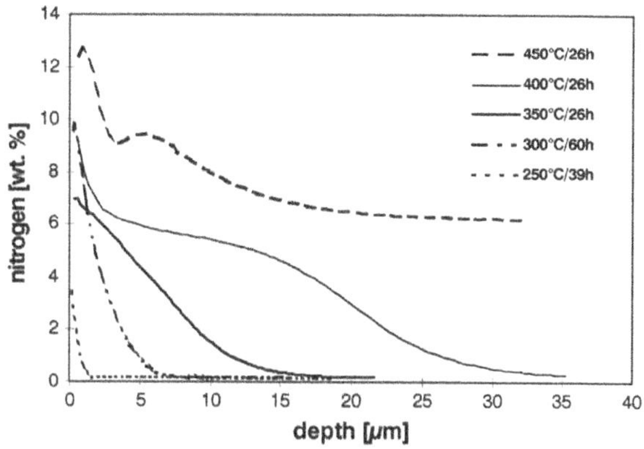

Fig. 3.9 Nitrogen depth profiles of plasma nitrided X6Cr17 samples [21]

Table 3.3 Effect of gas composition on growth rate [21]

Nitriding conditions: 350 °C, 16 h, 400 V	Growth rate: $W = d/\sqrt{t}$ (μm $h^{-0.5}$)	
N_2:H_2:Ar	X5CrNi18 10	X6Cr17
40:30:30	1.30	4.12
40:30:30 (450 V)	2.20	>5.95
60:10:30	1.18	3.85
60:30:10	1.31	3.30
80:10:10 (450 V)	0.50	3.05

The austenitic-ferritic steel showed the thinnest nitride layer (average 1.7 μm). The nitrided case in the X5CrNi18 10 sample was relatively thin (2.5 μm) and had a sharp interface with the core material, however, a thicker layer (14.6 μm) grown in the X6Cr17. The high internal stresses in the case resulted in needle-like structures, which were interpreted as deformation bands.

It was reported that the microstructure of low temperature plasma nitrided X8CrNiTi18 10 steel also consists of expanded austenite as observed in other stainless steels. With plasma nitriding of duplex steel below 350 °C, ferrite transformed into expanded austenite. This transformation corresponds for reduction of ferrite content in comparison with untreated sample. However, it should be noted that the behavior of ferritic phase was different in the ferritic steel X6Cr17 and the duplex steel X2CrNiMoN22 5 3. This difference rises from the presence of 4 wt% nickel in the ferritic phase of the duplex steel which aids the transformation into austenite.

Transformation of ferrite into expanded austenite at 350 °C, caused the reduction of ferritic phase. At higher nitriding temperatures, the transformation of substrate austenitic phase into ferrite and CrN occurred. As a result, the microstructure constituents were CrN, ferrite, and γ′-Fe_4N after nitriding at 500 °C for 20 h. Due to the thickness of nitrided layer, no substrate austenitic phase was observed.

Table 3.4 Thicknesses of compound layer and diffusion zone of nitrided samples [22]

Thickness (μm)	500 °C, 5 h	500 °C, 7 h	540 °C, 5 h	540 °C, 7 h	580 °C, 5 h	580 °C, 7 h
Compound layer	14	28	10	8	9	3
Diffusion layer	71	75	90	95	70	80

After plasma nitriding of ferritic steel at 350 °C for 20 h, the microstructure was constituted by ε-Fe_2N_{1-x} and ferrite. Prolonging the treatment duration to 40 h, led to an increase in the amount of ε-Fe_2N_{1-x} phase. However, it was reported that at higher temperatures, γ′-Fe_4N and CrN also were formed as well as existing ε-Fe_2N_{1-x}.

3.1.5 *Tool Steels*

Forati Rad et al. [22] observed the microstructure of plasma nitrided AISI H11 tool steel. The applied parameters were temperatures of 500, 540 and 580 °C, durations of 5 and 7 h with atmosphere of 75 % N_2 + 25 % H_2.

It was observed that, the compound layer and diffusion zone were distinguishable in the microstructure of the sample nitrided at 580 °C for 5 h. Table 3.4 presents the thickness of compound layer and diffusion layer for sample nitrided with different times and temperatures. As seen in Table 3.4, the thickness of both the diffusion zone and compound layer was influenced by the treatment time and temperature. Figure 3.10 shows SEM micrographs of samples. According to this figure, the average size of nitride nanoprecipitates for sample nitrided at 500 °C for 5 h was 118 nm which increased to 145 nm with increasing the treatment duration to 7 h.

Figure 3.10c, d show that at higher treatment temperature and/or time agglomeration of nanoprecipitates occurs. Sputtering on the sample surface enhanced in this atmosphere due to the high concentration of nitrogen ions and the presence of heavy nonionized nitrogen molecules in plasma media. Besides, the energy of nitrogen atoms was low and tended to recombine and form molecular N_2 on the sample surface. The combination of these effects together with the agglomeration of nitride precipitates led to the formation of micro-pores on the surface of the samples which appeared as dark points in the figure.

In the case of phase constituents, it was observed that, by increasing the time and/or temperature, the amount of ε-phase decreased, while that of γ′-phase increased.

These effects were also observed by Baki Karamiş et al. [23].

Basso et al. [24] also observed two distinct layers in the microstructure of plasma nitrided AISI H13 steel. They did plasma nitriding applying these parameters:

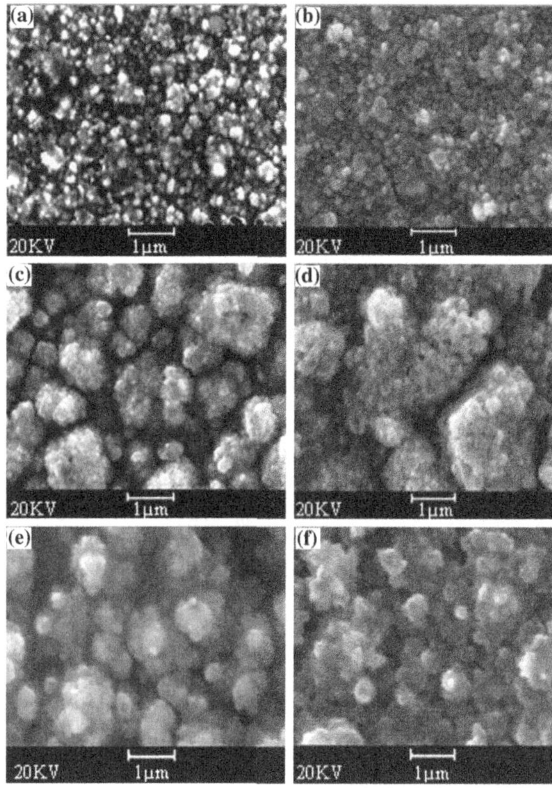

Fig. 3.10 SEM micrographs of the surface of the samples plasma nitrided in 75 % N_2 + 25 % H_2 atmosphere at: **a** 500 °C for 5 h, **b** 500 °C for 7 h, **c** 540 °C for 5 h, **d** 540 °C for 7 h, **e** 580 °C for 5 h, and **f** 580 °C for 7 h [22]

temperature of 360–520 °C, duration of 1–36 h, atmosphere of 20 % N_2 + 80 % H_2, pressure of 400 Pa, and voltage of 380 V.

Below 480 °C, a non-compact and above 480 °C, a compact compound layer were obtained for 4 h treatment.

For all samples, the ε-$Fe_{2\text{-}3}N$ phase was detected. However, for samples treated above 440 °C, the presence of γ′-Fe_4N was also reported. Due to the low diffusivity of N in low temperatures the formation of γ′-Fe-$_4$N was prohibited.

Two main differences were reported between treating at temperatures below 480 °C and above that: the formation of a compact compound layer and CrN phase.

At a fixed temperature, the thickness of nitrided layer formed in different durations is presented in Table 3.5. As can be seen, the layer thickness increased with time. Besides, even at low temperatures, the formation of γ′-Fe_4N was reported in long periods of time.

As shown in Fig. 3.11, the γ′-Fe_4N phase was detected in trace amounts when longer nitriding duration (36 h) was applied at 400 °C therefore, due to its low amount, ε-$Fe_{2-3}N$ is the main phase in the nitride layer.

Considering the results reported above, it was concluded that, different nitride layers can be obtained on the surface of AISI H13 tool steel by adjusting the plasma

Table 3.5 Nitride layer thicknesses for AISI H13 tool steel samples nitrided at 400 °C for different durations [24]

Nitriding time (h)	Layer thickness (μm)
1	25.2
4	36.1
9	44.0
16	50.0
36	66.8

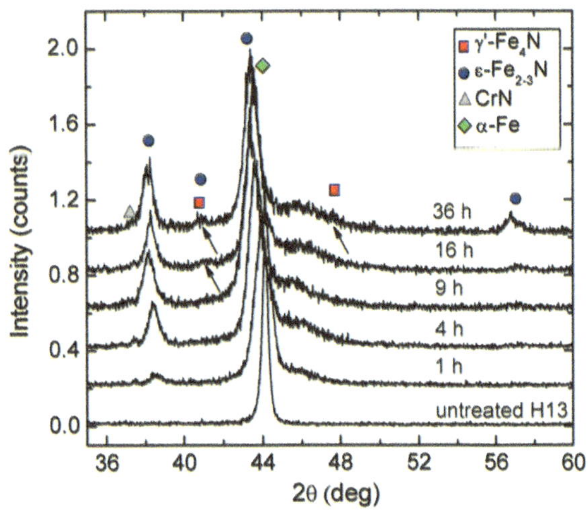

Fig. 3.11 XRD patterns of AISI H13 tool steel samples nitrided at 400 °C for different times [24]

nitriding temperature. Briefly, samples nitrided at T $\leq$ 400 °C exhibited a modified region near the surface whose structure consisted mainly of the ε-$Fe_{2-3}N$ phase. The formation of γ′-Fe_4N and CrN phases was favored above 440 °C. A clear microstructural rearrangement that led to the precipitation of ε-$Fe_{2-3}N$ and γ′-Fe_4N phases to form a compact compound layer, took place within the range of 480–520 °C.

Zagonel et al. [25] studied plasma nitriding of AISI H13 steel with following parameters: temperature of 260–510 °C, duration of 5 h, atmosphere of 30 % N_2 + 70 % H_2, pressure of 400 Pa and voltage of 380 V. It was reported that, the nitride layer formed at 260 °C only consisted of the original ferritic α phase. With increasing process temperature to 330 °C, the ε-$Fe_{2-3}N$ phase began to be formed. With a further process temperature increment to 370 and 420 °C, the presence of this phase in the material was consolidated. With further increase of temperature to 490 °C, the γ′-Fe_4N phase began to nucleate. This occurred as a result of fast diffusion of nitrogen and a nitrogen depletion concentration at the surface. Consequently at 510 °C this phase was also consolidated as well as ε-$Fe_{2-3}N$.

It was reported that temperature increment led to a less disorder in the structure owing to the enhancement of nitrogen mobility at high temperatures which promotes the formation of crystalline phases. On the other hand, applying lower nitriding temperatures led to the nitrogen accumulation on the top of the sample which in turn induces a disordered and stressed network.

Figure 3.12a, b shows the cross-section of the samples nitrided at 370 and 420 °C, respectively. Figure 3.12a shows a diffusion layer and a layer with a high nitrogen concentration at the grain boundary (which is seen as dark area) as well as few randomly distributed precipitates (which is seen as white points). The darker region at the grain boundary indicates that diffusion took place along these paths. Therefore, two main results could be obtained by low temperature plasma nitriding; (1) hindering nitride precipitation as a result of reduction in mobility of alloying elements; and (2) enhancing free nitrogen at the grain boundaries, preferring this diffusion path.

Soleimani et al. [26] plasma nitrided DIN 1.2210 cold work tool steel with these parameters: temperatures of 450, 500, and 550 °C, durations of 3, 6, and 9 h, atmosphere of 30 % N_2 + 70 % H_2 and pressure of 53 Pa.

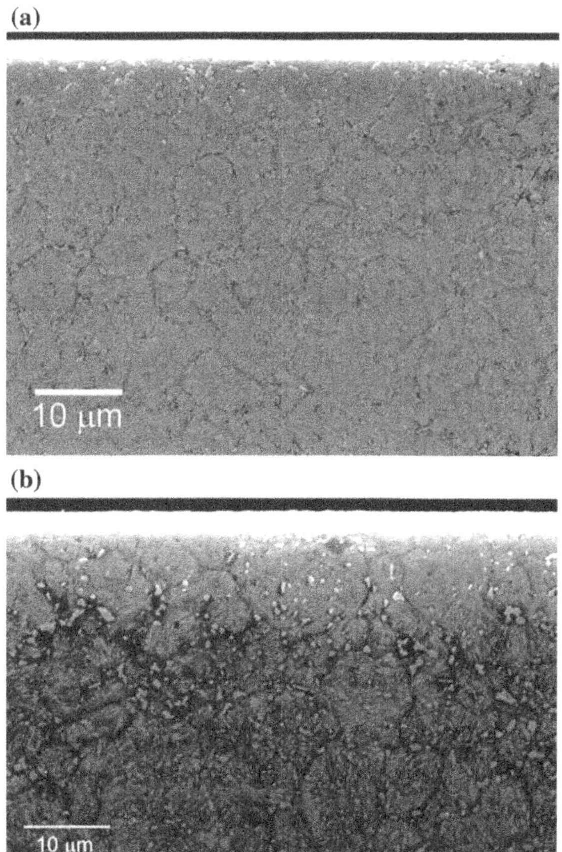

Fig. 3.12 **a** Micrograph of sample nitrided at 370 °C in which, the dark regions at the grain boundaries show the existence of nitrogen. **b** SEM image showing the structures of a sample nitrided at 420 °C [25]

It was observed that, plasma nitriding under different conditions led to the formation of the compound layers with thicknesses of 2–7 μm. The growth of the compound layer and diffusion zone is a diffusion-controlled phenomenon, therefore the thicknesses of these two layers increased with nitriding time and temperature, as expected. Increasing the nitriding temperature resulted in the enhancement of nitrogen diffusion coefficient and consequently, the formation of thicker regions.

It is obvious from Fig. 3.13 that the microstructure of nitrided samples consisted of ε and γ′ nitrides. Moreover, it can be seen that surface decarburization was enhanced with treatment temperature. Consequently, the intensity of the ε-Fe_3N phase diminished while the γ′-Fe_4N phase was intensified.

The elimination of iron peaks was also seen with treatment temperature at a constant duration of 6 h. This could be due to the layer thickening with increasing temperature and consequently preventing the incident X-ray from reaching the substrate.

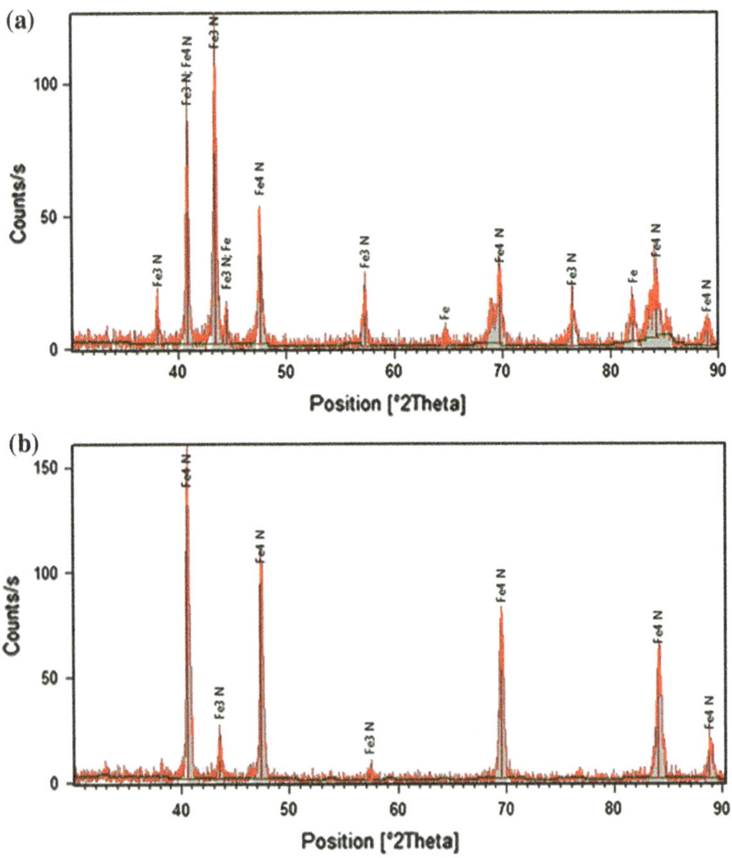

Fig. 3.13 XRD patterns of the samples plasma nitrided at **a** 450 °C and **b** 550 °C for 6 h [26]

Table 3.6 Nitrided layers depth on AISI D2 tool steel after plasma nitriding at 470 °C under different conditions [27]

$\%N_2$ in nitriding atmosphere	5			10			20		
Nitriding duration (h)	1	3	6	1	3	6	1	3	6
Nitrided layer depth (μm)	78	108	132	99	144	189	106	180	208

Conci et al. [27] investigated the effects of different amounts of nitrogen in the nitriding atmosphere on the microstructural and wear properties of AISI D2 tool steel. The applied parameters were: temperature of 470 °C, durations of 1, 3 and 6 h, pressure of 540 Pa, and atmospheres of 5, 10 and 20 vol% N_2 + H_2 (balance).

As can be seen in Table 3.6, nitrided depth increased by the nitrogen amount. This increment is significant from 5 to 10 % but slight from 10 to 20 %.

In case of using 5 % N_2 just α-Fe and CrN were detected as constituents of the surface. With increasing the nitrogen content to 10 %, $\varepsilon\text{-Fe}_{2\text{-}3}N$ was also detected as well as α-Fe and CrN. With further increasing of nitrogen to 20 %, the amount of $\varepsilon\text{-Fe}_{2\text{-}3}N$ increased and α-Fe and CrN were also detected.

Figure 3.14a–c shows the increasing of diffusion zone thickness and precipitation of nitrides with nitrogen potential. The increasing of nitrogen content in the atmosphere led to the further precipitation of nitrides either in amount and size as

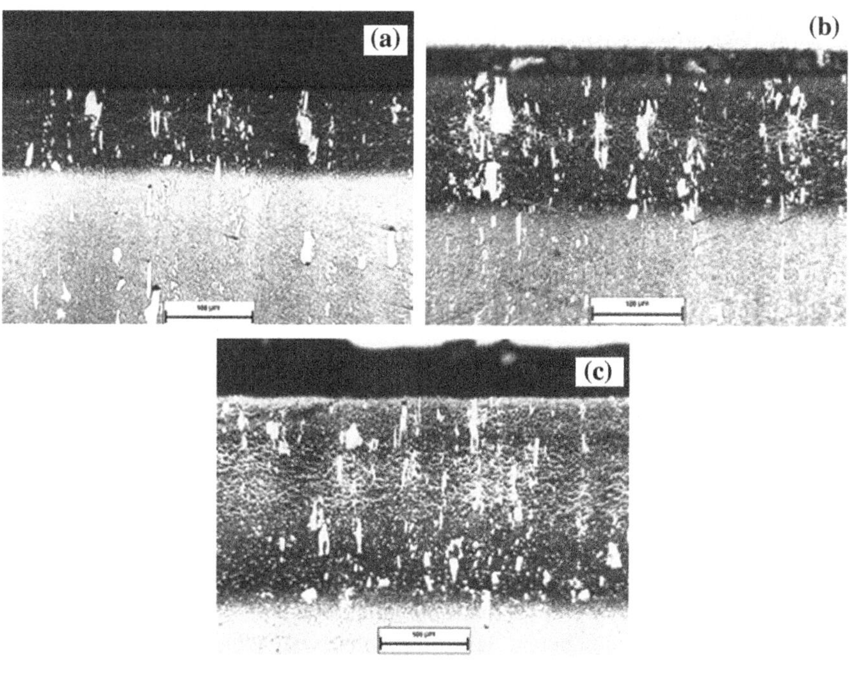

Fig. 3.14 Microstructures of the nitrided layers produced on the surface of AISI D2 tool steel after plasma nitriding at 470 °C for 3 h using nitrogen potentials of **a** 5 %, **b** 10 %, and **c** 20 % [27]

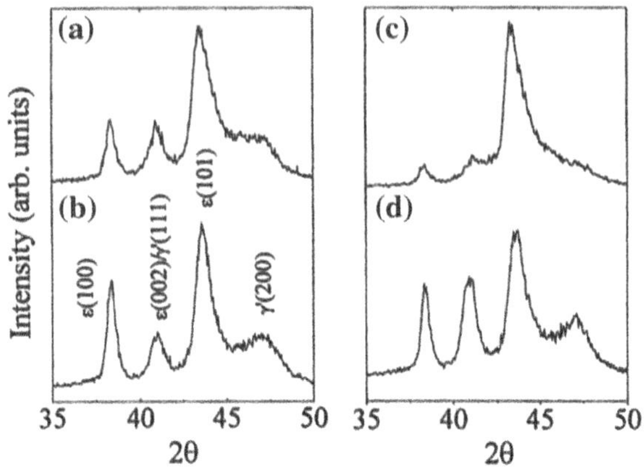

Fig. 3.15 XRD patterns of the samples pulsed DC nitrided for **a** 1 h, **b** 3 h; conventional DC nitrided for **c** 1 h and **d** 3 h [28]

can be seen in Fig. 3.14c. Also, using 20 % N_2 in nitriding atmosphere led to the formation of compound layer.

Miola et al. [28] plasma nitrided AISI H12 tool steel samples with two conventional and pulsed DC methods and compared the results. Treatments were done at 500 °C, in atmosphere of 80 % H_2 + 20 % N_2, under pressure of 600 Pa for 1–6 h.

For samples nitrided with both methods, a compound layer mainly consisted of γ' and ε was detected as seen in XRD patterns in Fig. 3.15. For sample nitrided by DC glow discharge for 1 h, no γ' phase was detected but appeared with increasing treatment time. This can be seen clearly in Fig. 3.15.

In case of conventional DC plasma nitriding, the thickness of the hardened layer was measured about 38 μm for the sample nitrided for 2 h and about 85 μm for that nitrided for 6 h.

Kinetics of modified layer's growth for both methods are shown in Fig. 3.16. Sample nitrided with conventional DC plasma followed parabolic law with saturation limit for duration higher than 5 h. However for puled DC nitriding, two growth regimes were followed by the layer. Faster growth was seen for initial stage and slower for final stage. Also a saturation was seen for durations higher than 6 h.

3.1.6 Constructional and Structural Steels

Corengia et al. [29] investigated the plasma nitriding of AISI 4140 low alloy steel with applying these parameters: temperature of 500 °C, durations of 1, 2, 4, 15, 20 and 28 h, atmosphere of 25 % N_2 + 75 % H_2, pressure of 60 Pa, and voltage of 750 V.

The thickening of compound layer was observed with nitriding time. This thickness increment was very fast in the first hours of treatment but then the growth rate diminished. This can be explained by sputtering which caused the elimination

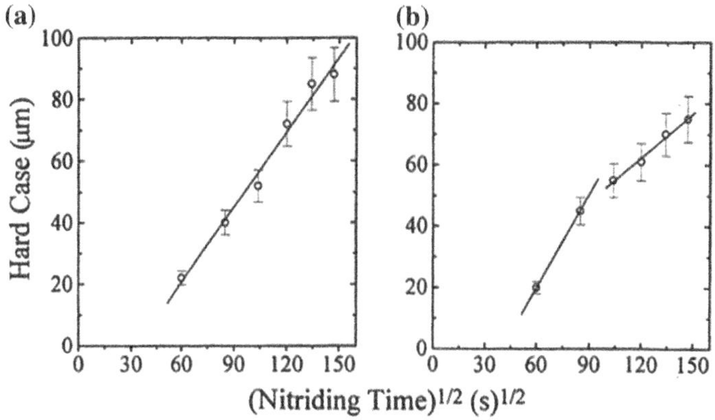

Fig. 3.16 Thickness of the hardest layers versus square root of the nitriding time for **a** conventional plasma nitrided and **b** pulsed DC nitrided samples [28]

of the formed nitrides, the denitriding and the role of the layer as a barrier for the formation of free Fe atoms and N atoms diffusion.

For short treatment times, ε-$Fe_{2\text{-}3}N$ and γ′-Fe_4N were present in the microstructure of nitride layer. However, for processing times longer than 15 h, the ε phase was not detected. Therefore, it can be seen also from Table 3.7 that, for durations longer than 15 h, just γ′-Fe_4N was formed.

Higher depth values of the diffusion zone were obtained with increasing of the treatment time.

Ochoa et al. [30] studied several samples to achieve the best condition for plasma nitriding of CK45 steel used in steel gears. The applied parameters for plasma nitriding were: temperature of 540 °C, duration of 4 h, atmosphere of 70 % N_2 + 30 % H_2, pressure of 500 Pa, and voltage of 500 V. These conditions led to the formation of compact, uniform, and homogeneously nitrided layers as shown in Fig. 3.17.

Figure 3.18 shows the intensity of nitride phases and nitrogen concentration along the compound layer. As observed in this figure, the γ′-Fe_4N and ε-$Fe_{2\text{-}3}N$ phases are diminishing along the compound layer while the α phase increases. Indeed, at a depth of ~20 μm, the presence of ε and γ′ phases are negligible indicating the incorporation of nitrogen in the matrix of the α phase.

Table 3.7 Diffraction intensities of the ε and γ′ phases for different nitriding durations [29]

Nitriding duration	γ′-Fe_4N	ε-$Fe_{2\text{-}3}N$
1	74	26
2	76	24
4	84	16
15	93	7
20	95	5
28	96	4

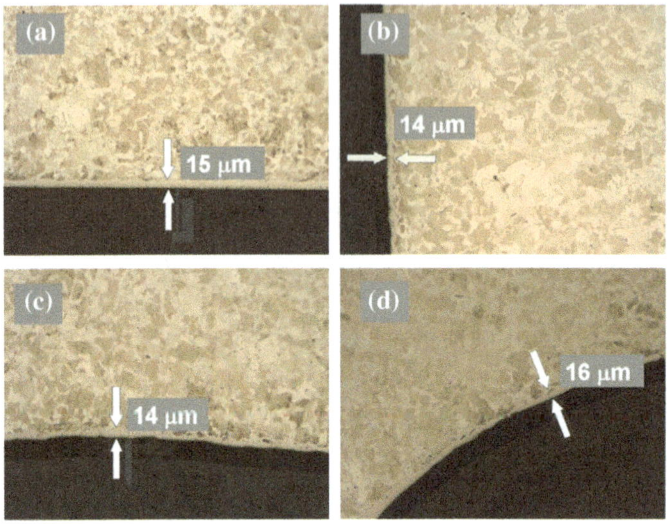

Fig. 3.17 **a** and **b** Nitrided layer at the top and flank part of the teeth, respectively; **c** and **d** Nitrided layer details in two different regions of the bottom part of the teeth [30]

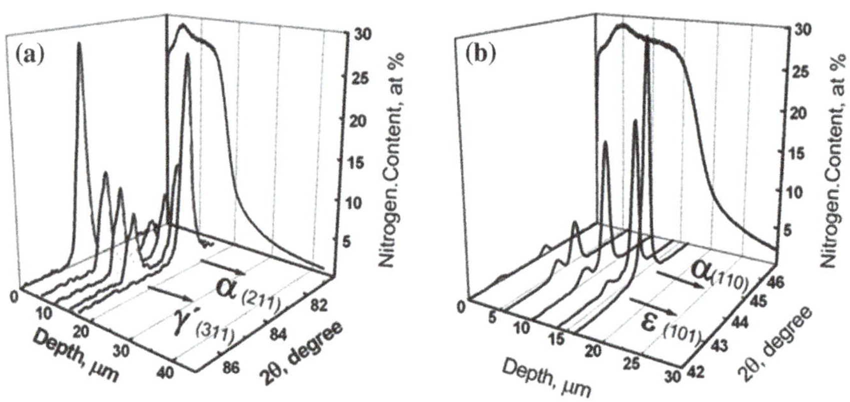

Fig. 3.18 Phase composition of compound layer with thickness (for sample displayed in Fig. 3.17a). The diffractogram is divided in reflections austenitic associates with the a) γ', α and b) ε, α crystalline phases [30]

Li et al. [31] observed the "edge effect" (reported in Chap. 2, Sect. 2.1.1 by Olzon-Dionysio et al. [7] about AISI 316L austenitic stainless steel) in the 42CrMo low alloy steel. They plasma nitrided the samples at temperatures of 450 and 540 °C, duration of 4 h, atmosphere of NH_3, pressure of 500-600 Pa, and voltage of 500–700 V.

Figure 3.19a shows that the edges of the cathodic plasma nitrided sample had different colors compared to the central area. This phenomenon is known as edge

Fig. 3.19 Appearance of 42CrMo steel samples nitrided at **a** cathodic and **b** anodic potential [31]

effect which is a common problem with the DC plasma nitrided samples at cathodic potential. It is occurred as result of nonuniform sputtering and deposition during the nitriding process. In contrast, anodic plasma nitriding produced a uniform silvery color in the whole surface of the samples as seen in Fig. 3.19b.

It was reported that, the surface of sample nitrided with anodic potential consisted of small grains. For comparison a glass substrate was treated with same conditions and showed the same morphology. Therefore it was concluded that, these small grains were deposits from cathode sputtering. But very few small grains were observed on the surface of samples nitrided with cathodic potential.

The thickness of compound layer was about 4–5 μm at 450 °C and 8-10 μm at 540 °C for both potentials.

For the sample treated at 450 °C for 4 h, nitride phases were detected as well as disappearing some small α-Fe peaks. During plasma nitriding at 540 °C for 4 h for both potentials, γ'-Fe_4N and ε-$Fe_{2\text{-}3}N$ phases were formed.

Basu et al. [32] investigated the plasma nitriding of SAE 52100 low alloy–high carbon steel with following parameters: temperatures of 450, 500, and 560 °C, durations of 1, 3, and 5 h, atmosphere of 35 % N_2 + 65 % H_2, and voltage of 540–710 V.

As can be seen in Fig. 3.20a, in the sample treated at 450 °C, only weak γ'-Fe_4N peaks were present as well as the strong α-Fe peaks. However, plasma nitriding at 500 and 560 °C resulted in the formation of ε-$Fe_{2\text{-}3}N$ and γ'-Fe_4N nitrides along with α-Fe matrix. At a given temperature, the relative intensity of Fe_xN peaks increased with an increase in nitriding time. It can also be observed from the micrograph in Fig. 3.20b that the nitrides produced were very fine in nature and uniformly distributed on the surface of the treated steel.

It was reported that, nitride volume fraction increased with nitriding time at a constant nitriding temperature (Fig. 3.21). Furthermore, nitride volume fraction was also increased with nitriding temperature. Nitriding at 560 °C for 5 h resulted in the formation of the maximum nitride volume fraction (94.5 %).

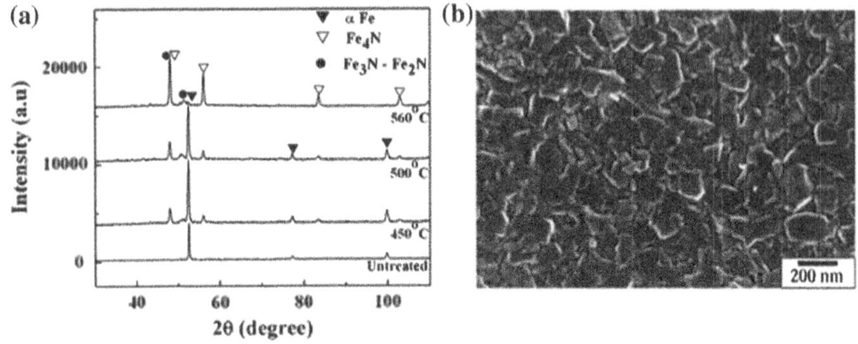

Fig. 3.20 **a** XRD patterns of the sample nitrided at different temperatures for 3 h. **b** High resolution micrographs showing plasma nitrided surface of plasma nitrided steel after nitriding at 560 °C for 5 h [32]

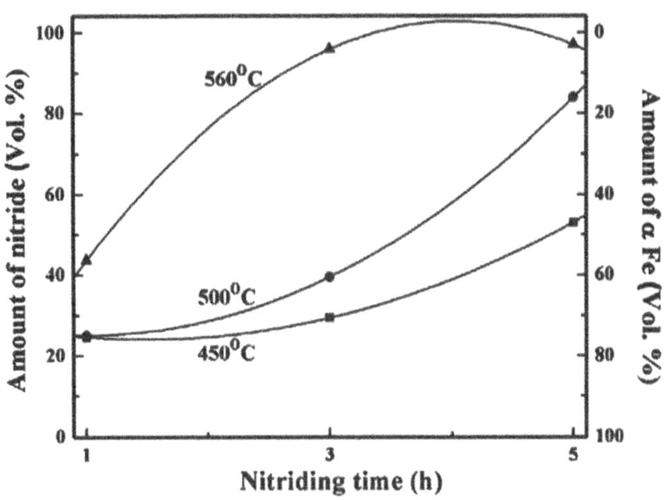

Fig. 3.21 Amount of nitrides versus nitriding time at different temperatures [32]

Podgornik et al. [4] compared AISI 4140 (42CrMo4) and AISI A355 (34CrAlMo5) plasma nitrided with following parameters: temperature of 540 °C, duration of 17 h, atmosphere of 0.6 % N_2 + 99.4 % H_2, and pressure of 254 Pa.

It was reported that for both steels nitrided under above-mentioned conditions, a very thin compound layer of <1 μm was formed on top of a diffusion zone of ~0.3 mm. A comparison of the sample microstructures after conventional plasma and pulsed plasma nitriding showed completely identical microstructures, as well as compound layer thicknesses for both the steels used in this investigation.

Li et al. [33] investigated the effects of post-oxidation treatment on properties of pulsed DC plasma nitrided AISI 4140 steel. Plasma nitriding was done at 540 °C

Table 3.8 Compound and oxide layer thickness of plasma nitrided and post-oxidized AISI 4140 steel [33]

Treatments	Compound layer (μm)	Oxide layer (μm)
Plasma nitriding	8–10	–
Plasma nitriding + oxidizing at 500 °C	7–8	2–2.5
Plasma nitriding + oxidizing at 450 °C	8–9	1.2–1.5
Plasma nitriding + oxidizing at 400 °C	8–9	0.8–1
Plasma nitriding + oxidizing at 350 °C	8–10	0.2–0.4

for 4 h under a pressure in the range of 500–600 Pa and in atmosphere of pure ammonia (NH_3). Post-oxidizing treatments were performed at temperatures of 350, 400, 450, and 500 °C for 2 h under pressure of 120 Pa.

The compound layer formed on the plasma nitrided sample was 8–10 μm thick and constituted by γ'-Fe_4N and ε-$Fe_{2\text{-}3}N$. As presented in Table 3.8, increasing of oxidizing temperature resulted in the reduction in compound layer thickness and increasing of the oxide layer thickness. This could be due to the diffusion of nitrogen inward into the substrate or diffusion outward and accumulation of nitrogen in the interface of nitride and oxide layers.

The surface layer of plasma post-oxidized samples was composed of Fe_2O_3, Fe_3O_4, iron nitrides, and α-Fe. An increase in the oxidation temperature led to further detection of oxides and α-Fe phases. The amount of iron oxides Fe_2O_3 and Fe_3O_4 was dependent on the post-oxidizing temperature as the samples post-oxidized at 400 and 450 °C showed the highest amount of Fe_3O_4.

Ebrahimi et al. [34] investigated the effects of plasma oxidation after nitriding on properties of AISI 4140. Plasma nitriding was done at temperatures of 530, 570 and 630 °C for 5 h in atmosphere of 80 % N_2 +20 % H_2 under a pressure of 4 mbar. Post-plasma oxidizing treatment was done at 450 °C for 1 h in atmosphere of 50 % H_2 + 50 % O_2.

It was observed that the compound layer thickness increased with temperature due to more nitrogen diffusion at higher temperatures (Table 3.9). Post-oxidizing treatment resulted in the formation of an oxide layer on top of nitride layer as seen in Fig. 3.22. After oxidation treatment, the thickness of compound layer decreased. This was due to sputtering effect on compound layer and its decomposition. Consequently the thickness of compound layer decreased and iron atoms reacted with oxygen to form oxide layer. It was observed that nitriding at 630 °C produced an intermediate layer between compound layer and diffusion layer called transformed austenite layer. This layer was formed when the nitriding temperature exceeds 590 °C, due to the austenite stabilization effect of nitrogen. The transformations of austenite to α and γ' also happened during slow cooling from temperatures above 590 °C. It can be seen in Table 3.9, the thickness of transformed austenite layer was increased with oxidation. This occurred due to the releasing of nitrogen atoms during oxidation and their diffusion inward to thicken the transformed austenite layer.

Table 3.9 The thicknesses of compound layer, transformed austenite layer, and oxide layer and surface roughnesses of the samples that were plasma nitrided before and after oxidation [34]

Specimen	Applied voltage (V)	Compound layer thickness (μm)	Austenite layer thickness (μm)	Oxide layer thickness (μm)	Surface roughness (μm)
Untreated	–	–	–	–	0.05
Nitrided at 530 °C	715	6.95	–	–	0.13
Nitrided at 570 °C	800	13	–	–	0.17
Nitrided at 630 °C	850	18.2	3.5	–	0.25
Nitrided at 530 °C—Oxidized	730	5.7	–	2.1	0.09
Nitrided at 570 °C—Oxidized	745	11.9	–	3.2	0.11
Nitrided at 630 °C--Oxidized	750	17.7	4.3	3.9	0.20

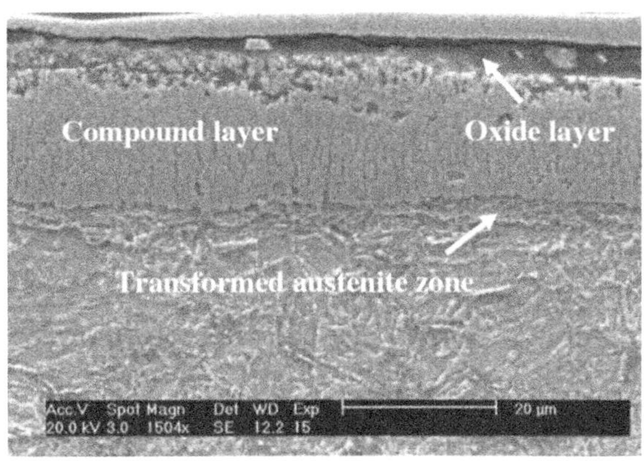

Fig. 3.22 SEM micrograph of the sample nitrided at 630 °C after oxidation at 450 °C [34]

As seen in Table 3.9, increasing treatment temperature also led to an increase in surface roughness. This happened due to the requirement to higher voltage with increasing temperature so the sputtering intensity increased and the surface was attacked with more energetic ions. Increasing the roughness after plasma post-oxidation treatment was attributed to covering of surface and its pores with a compact and dense oxide layer.

The compound layers of samples plasma nitrided at 530, 570, and 630 °C, were mainly consisted of ε and γ′ nitrides. For 570 °C, ε was dominant with traces of γ′

however for 630 °C, γ′ became predominant. This was owing to the fact that at that temperature ε phase was formed but with slow cooling transformed into γ′.

The oxide layer of oxidized samples was mainly constituted by Fe_2O_3 and Fe_3O_4. The samples nitrided at higher temperatures of 570 and 630 °C showed more Fe_3O_4 phase because of larger surface of ε compared to γ′ which resulted in more oxygen uptake.

3.2 Mechanical Properties

3.2.1 Austenitic Stainless Steels

Liang [5] (see Sect. 3.1.1) observed an ascending temperature-dependent behavior for hardness of plasma nitrided AISI 304 austenitic stainless steel as can be implied from Fig. 3.23. This is because of the deeper nitrogen penetration and thicker layer formation.

Due to the inducing compressive residual stress in nitrided layers, the calculation of this stress could be important from a viewpoint of hardness. Because, compressive residual stress could play a great roll in enhancing the surface hardness of nitrided layers. Using Stoney equation (Eq. 3.1), the calculation of the average induced stress in thin layers was possible. For this purpose, the deflection of the free end of a cantilevered beam type substrate was measured to determine its radius of curvature. Stoney equation is as follows:

$$\sigma_f = \frac{E_s d_s^2}{6R}(1 - \vartheta_s)d_f \tag{3.1}$$

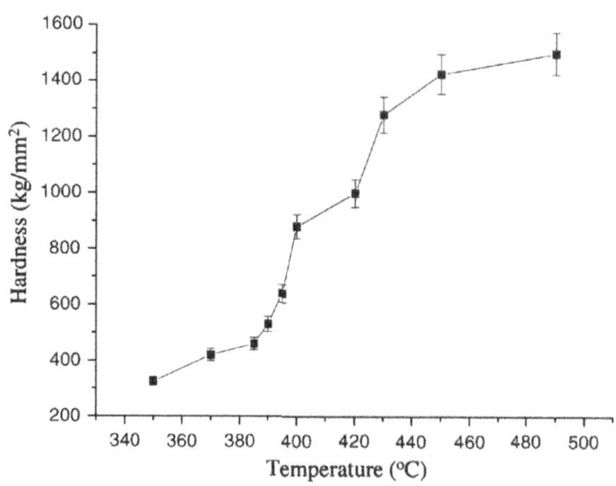

Fig. 3.23 Surface microhardness versus processing temperature [5]

Table 3.10 The results of residual stresses in nitrided layer [5]

Processing temperature (°C)	380	400	420	450	480
Thickness of nitrided layer (μm)	3.5	4.7	6.8	12.3	25
Residual stress (MPa)	622	895	1021	886	691

where E_s and ϑ_s are Young's modulus and Poisson's ratio of substrate, d_s and d_f are the substrate and layer thickness, respectively, and R is the radius of curvature of cantilever after treatment. The stress value calculated using above equation is shown in Table 3.10.

Singh et al. [6] (see Sect. 3.1.1) observed an increase in roughness after plasma nitriding of AISI 304 austenitic stainless steel (Fig. 3.24). The high sputtering rate corresponded for the increased surface roughness when nitriding in the atmosphere with higher hydrogen content. Furthermore, the increasing discharge voltage was the other factor for surface roughness, which occurs with reducing the nitrogen content in the nitriding atmosphere. It was concluded in this study that, the dependence of surface roughness on the composition of nitriding atmosphere is more severe than nitriding time and temperature.

For all mechanically prepared samples, surface hardness improvement was observed with increasing nitrogen content in the treatment atmosphere which can be attributed to the higher amount of nitrides. The highest hardness value belonged to the mirror polished sample nitrided with higher nitrogen concentration.

Sharma et al. [7] (see Sect. 3.1.1) found no increase in surface hardness of the samples with nitriding in pure N_2 atmosphere. It was observed that when the sample was nitrided at 540 °C for 4 h in N_2–H_2 plasma under pressure of 3.64 torr, surface hardness increased away from the center of the sample. After nitriding with these conditions, the surface hardness increased up to three to four times toward the periphery of the samples (both AISI 302 and 304 samples). To obtain a uniform hardness over the whole surface of these samples, the nitriding time might have to be more than 4 h. However, for the AISI 302 sample nitrided at relatively higher

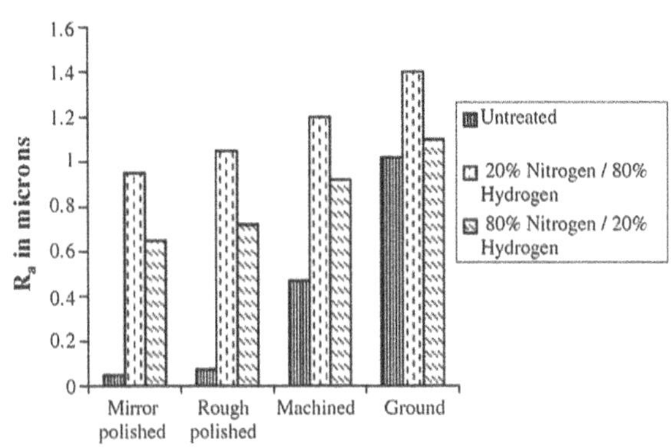

Fig. 3.24 Surface roughness of all the four mechanically prepared untreated and plasma nitrided AISI 304 samples with different gas compositions [6]

Table 3.11 Maximum surface hardness of AISI 302 and AISI 304 samples after plasma treatment at different discharge conditions [7]

Sample	Pressure (torr)	Sample temperature (°C)	Discharge voltage (V)	Maximum surface hardness
AISI 304	3.6	540	580	1221
AISI 302	5.3	540	580	842
AISI 302	5.7	830	779	657

nitriding temperature (830 °C) for 2 h, surface hardness was measured at four different positions along the surface which were found to be more or less uniform (583, 578, 657, and 535 HV). It was observed that plasma nitriding at higher temperature for relatively less time led to a more uniform hardness. Table 3.11 presents the list of hardness values obtained at different plasma conditions in N_2–H_2 mixture.

At 700 °C and above, nitriding occurred solely by a diffusion-controlled process and was more dominant whereas between 550 and 650 °C, physical effects such as fracture of surface layers were predominant. That may be a reason why at 540 °C uniform hardness was not obtained. At higher temperature, generally above the optimum value of 500 °C, grains over the surface lost their strength and thus hardness decreased at 830 °C compared to that of 540 °C.

Gontijo et al. [8] (see Sect. 3.1.1) observed that increasing temperature resulted in higher hardness values for plasma nitrided AISI 304L austenitic stainless steel. This can be seen in Table 3.12. For the sample nitrided at 350 °C, with a layer thickness of 2.6 μm, no microhardness value was obtained. The matrix had a microhardness value of 220 HV. Expanded austenite and CrN were formed on the samples nitrided at 450 °C (Fig. 3.2c) and 500 °C (Fig. 3.2d). The s-phase (expanded austenite) layer got thicker with nitriding temperature. It was reported that, the presence of CrN led to the higher surface microhardness values for layers of the samples nitrided at 450 and 500 °C.

Menthe et al. [9] (see Sect. 3.1.1) observed the increment of hardness of plasma nitrided AISI 304L austenitic stainless steel with increasing nitrogen partial pressure.

A high adhesion strength was observed between modified layer and the substrate in bending test. Even with performing the test at −192 °C, the substrate material was still ductile. The layer itself was very compact with good adhesion to the substrate material.

Table 3.12 Microhardness and corresponding layer thickness values obtained at different nitriding temperatures [8]

Temperature (°C)	Microhardness (HV)	Thickness (μm)
350	–	2.6
400	1000	6.5
450	1145	14.6
500	1487	36.4

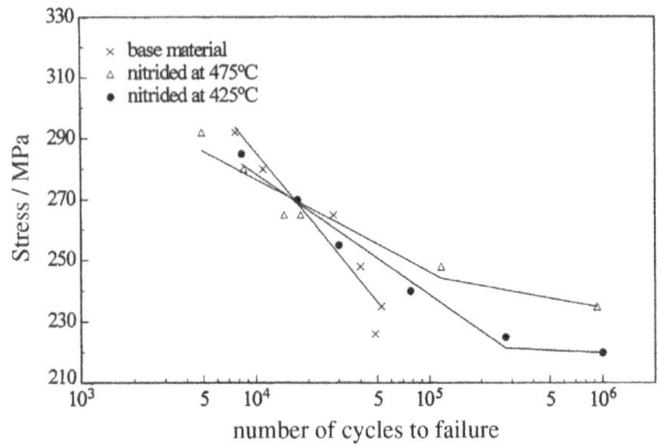

Fig. 3.25 S–N curves for unnitrided and plasma nitrided austenitic stainless steel [9]

As seen in Fig. 3.25, in the low cycle fatigue (LCF) range ($N_f < 50{,}000$) plasma nitriding did not have any apparent impact, but in high cycle fatigue (HCF), increased fatigue life was observed. This was due to the fact that in the case of HCF most of the fatigue life was spent during the crack initiation whereas in the LCF range, fatigue life was controlled by crack growth in the bulk material. Therefore, since the bulk material was not affected by plasma nitriding, the fatigue life during LCF was not changed significantly.

It was reported that, plasma nitriding strongly reduced the wear rate of samples, even under higher loads. The friction coefficient of plasma nitrided austenitic steel against ferritic or austenitic steel was lower compared to the unnitrided sample.

Jeong et al. [10] (see Sect. 3.1.1) observed that hardness of plasma nitrided AISI 316L increased with temperature. The reported hardness was in the range of 1115–1160 HV and 1248–1315 HV for treatment temperatures of 400 and 500 °C, respectively. This could be because of different phases formed at different temperatures (Fig. 3.3). Expanded austenite phase was dominant at lower temperatures but with increasing temperature, CrN got predominant. Even if the hardness of the s-phase was lower than that of the dual phase nitride (CrN and Fe_4N), due to the high nitrogen concentration, expanded austenite phase had compressive stress and lattice distortion and consequently high hardness and good wear resistance.

As seen from AFM images in Fig. 3.26, the grain size for the sample nitrided at 500 °C was larger than that of the sample nitrided at 400 °C. The concentration of nitrogen on the nitrided surfaces was high, therefore, the peak-to-valley height of the surface features was of the order of 1000 and 2000 nm for samples treated at 400 and 500 °C, respectively. This difference may be ascribed to the intense compressive strain.

The surface roughness (R_a) of samples was measured for different pulse frequencies and at two temperatures. It was reported that, R_a increased with frequency. Also, it was observed that the rate of roughness increment at 500 °C was larger than 400 °C. This was due to the higher sputtering rate in higher temperature.

Li et al. [12] (see Sect. 3.1.1) observed the hardness uniformity in pulsed DC plasma nitrided AISI 316L samples whereas in the conventional plasma nitrided

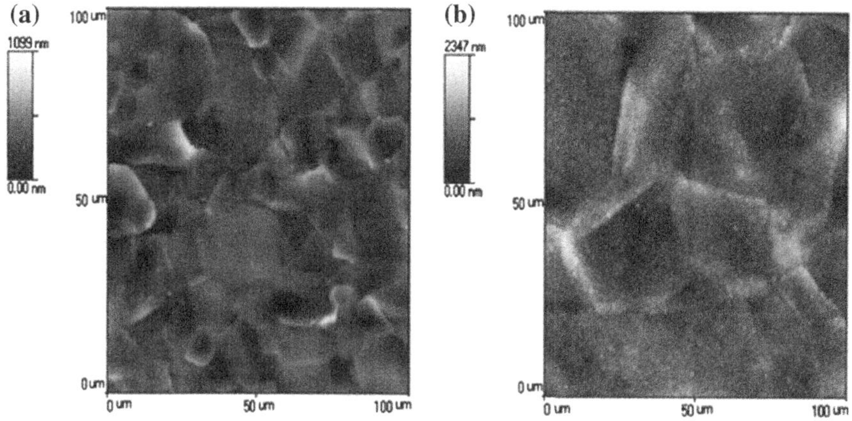

Fig. 3.26 AFM images from surface of the samples nitrided at: **a** 400 °C and **b** 500 °C [10]

samples, different hardness values were measured in the central region and the outer rings. In the cathodic nitriding process, the bombardment of the sample's surface was done by the ionized species. Furthermore, the distortions in the electric field around the corners and edges significantly affected the intensity of bombardment which led to the nonuniformity of the nitrided layer. However, in anodic nitriding process, the surface bombardment was done only by electrons. Also, the voltage drop in the anodic process was lower than cathodic process, because of concentration of the voltage drop of the inter electrodes on the cathodic sheath. Therefore, the energy of electrons bombardment was much weaker than that of ion bombardment.

It was also observed the increment of hardness with increasing treatment temperature as reported by other authors like Gontijo et al. [8] and Liang [5]. A 9 μm nitrided layer with hardness of 1200 $HV_{0.01}$ was achieved on the sample nitrided at 480 °C. For the samples nitrided at 540 °C, the thicker layer of about 36 μm was achieved with a constant microhardness of 133–1500 $HV_{0.01}$ along the layer.

Figure 3.27 shows the friction coefficients of the untreated and nitrided sample slide against AISI 52100 under a load of 10 N. As clearly seen in this figure, friction coefficient of the sample nitrided at 450 °C was lower and smoother compared to the unnitrided sample. However, nitriding at 540 °C did not cause any significant change in the average friction coefficient.

Figure 3.28 shows SEM images of the samples after wear test. The improvement of wear resistance is obvious. As seen in Fig. 3.28a and b, a deep and wide wear track with large plastic deformation as well as metallic particle adhesion on the track was clearly observed. After nitriding at 450 °C, the wear track became narrower and shallower compared to untreated sample. Besides, the formation of the relative dark and smooth oxidized pitch as well as the morphologies of the wear track indicated the domination of oxidation wear and slight abrasion in the wear process of this sample. By increasing the nitriding temperature, the wear track

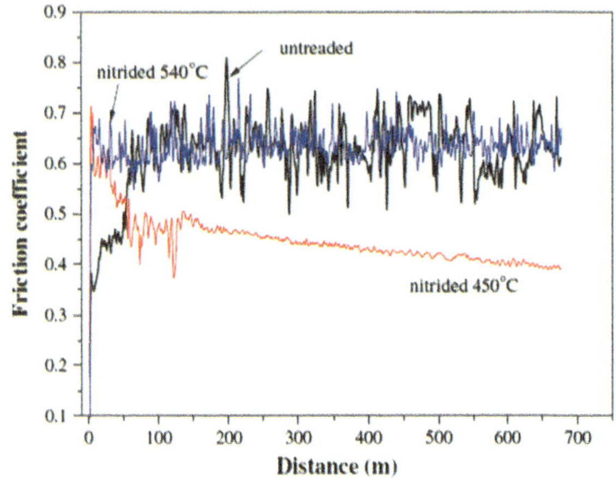

Fig. 3.27 The variations of friction coefficient with sliding distance for untreated and nitrided samples [12]

further decreased. The wear resistance of this sample was reported to be the maximum among all the samples.

The improvement of wear resistance was also reported by Xia et al. in the case of 1Cr18Ni9Ti austenitic stainless steel [13] (see Sect. 3.1.1). They used lubricant with different additives and observed the improvement of wear resistance in all cases. They also concluded that tribochemical reactions occurred between the stainless steel and the active elements of the additive during the sliding of the stainless steel disks against SAE52100 steel under the lubrication of liquid paraffin-containing sulphurized olefin. The resulting surface-protective films together with the adsorbed boundary lubricating film corresponded for the significantly reduced friction and wear.

3.2.2 *Martensitic Stainless Steels*

Figure 3.29 shows microhardness profiles of annealed and quenched-tempered pulsed DC plasma nitrided AISI 410 martensitic stainless steel samples at 350, 400, and 500 °C for 20 h [14] (see Sect. 3.1.2). All samples showed high surface microhardness values of above 1000 HV which dropped suddenly to the substrate microhardness values at the interface of case and core. It can be observed in Fig. 3.29 that higher surface hardness values were obtained with nitriding at temperatures of 350 and 400 °C.

A sharp change in hardness profile from case to core was also observed by Corengia et al. [15] in plasma nitrided AISI 410 martensitic stainless steel. The parameters used for plasma nitriding were: temperature of 400 °C, duration of 20 h, atmosphere of 25 % N_2 + 75 % H_2, pressure of 6 hPa, and voltage of 750 V.

According to Fig. 3.30, for all experiments, the mean friction coefficient of the nitrided sample was always lower than that of untreated samples.

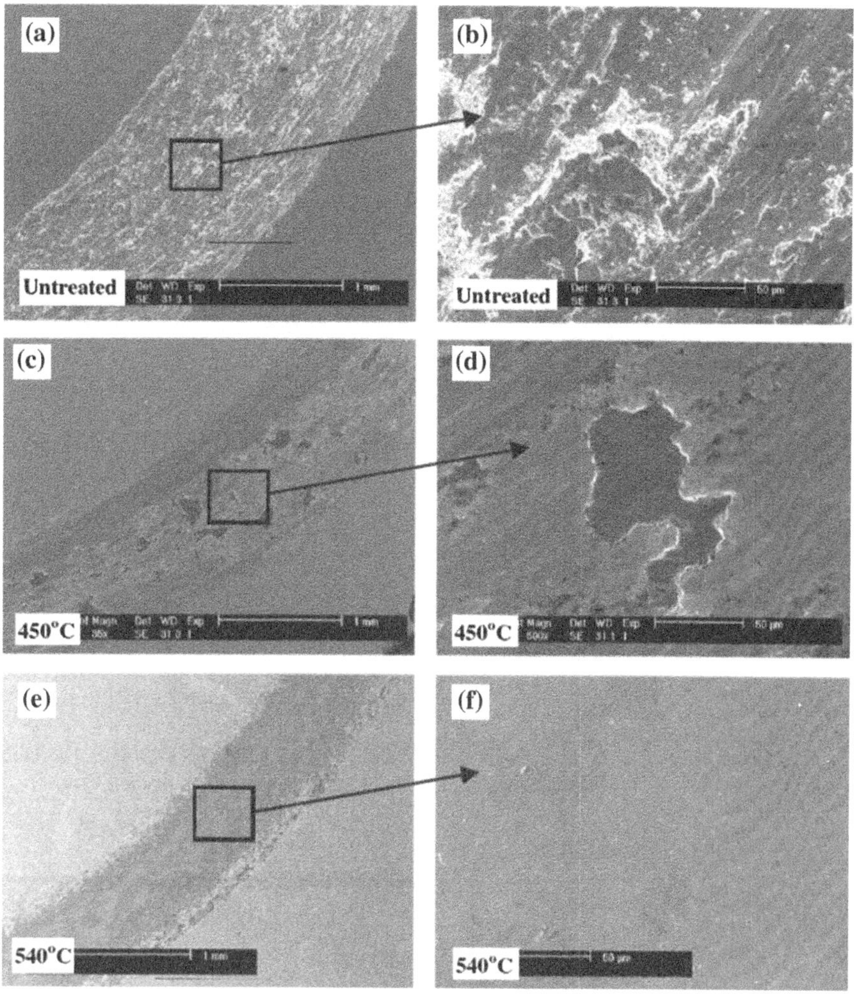

Fig. 3.28 SEM micrographs of wear tracks on the surface of samples **a** and **b** untreated, **c** and **d** nitrided at 450 °C, and **e** and **f** nitrided at 540 °C [12]

As seen in this figure, for untreated sample, very severe fluctuations were observed due to a stick-slip phenomenon. This phenomenon occurred between the asperities in contact. These were resulted by the difference between the static and kinetic friction coefficients.

However for plasma nitrided samples, two distinct regions were distinguishable (Fig. 3.30). In the first stage, very stable and smooth behavior was seen but thereafter fluctuations begin to appear. This phenomenon explained by production of hard debris as a result of spalling of the nitrided case. Consequently the appearance of the surface became rougher because of gradual removal of the nitrided case.

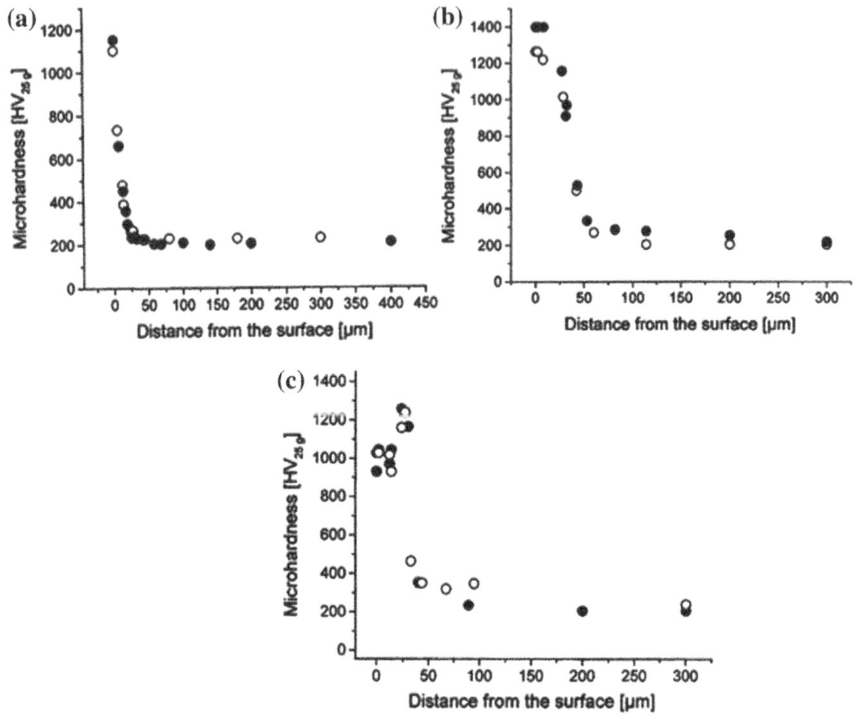

Fig. 3.29 Microhardness profiles of the pulsed DC plasma nitrided samples for 20 h at: **a** 350 °C, **b** 400 °C and **c** 500 °C. Quenched-tempered: black circle and annealed: white circle [14]

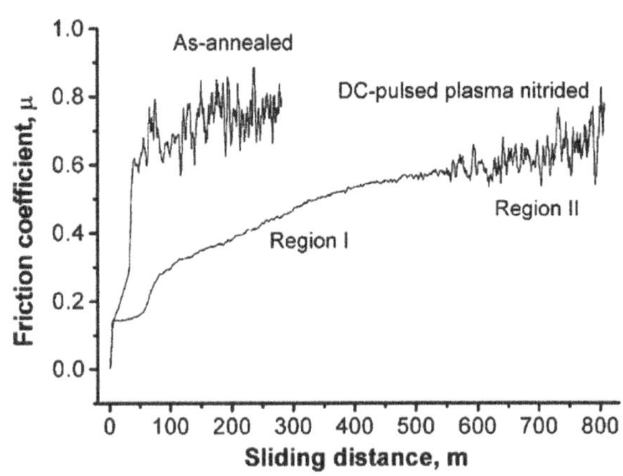

Fig. 3.30 Variations of friction coefficient for unnitrided and pulsed DC plasma nitrided samples [15]

They observed that a combination of different mechanisms was involved in the wear of unnitrided and nitrided samples. Unnitrided steel was degraded by abrasion, delamination, and the activation of tribochemical reactions, leading to surface and internal oxidation of the worn samples. Instead, oxidation and delamination were the main mechanisms dominating wear of the nitrided specimen under low applied loads. The hard nitrided case provided a more rigid substrate for the oxide films, and thus the contact pairs were typical oxide to oxide.

Yun-tao et al. [16] (see Sect. 3.1.2) observed the increment of hardness with decreasing temperature as noted in the case of conventional DC plasma nitrided martensitic stainless steel. It was reported that the microhardness of the substrate was 289 $HK_{0.025}$, while those of nitrided layers were higher than 1000 $HK_{0.025}$. The microhardness of the nitrided layer for the sample nitrided at 350 °C was about 4.3 times that of the substrate.

For the sample nitrided at 350 °C, the maximum residual stress was −620 MPa on the surface. However for the sample nitrided at 550 °C, this value was only 31 % of that of the sample nitrided at 350 °C. The compressive residual stress was resulted by the solid solution of nitrogen and the distortion of lattice in the nitrided layer. At 350 °C, the existence of α_N resulted in higher compressive residual stress (Fig. 3.31).

The ratio of erosion rate of the unnitrided sample, the 350 °C nitrided sample, and the 550 °C nitrided sample was 2.8:1:1.8.

Lower erosion rate was observed for samples nitrided at both temperatures compared to the unnitrided sample. This improvement of the erosion resistance of AISI 420 martensitic stainless steel after plasma nitriding can be ascribed to several factors such as the microstructure, the high surface hardness, and the high compressive residual stress in the nitrided layer.

The inverse relation of hardness and treatment temperature also was reported in the case of AISI 420 by Yun-tao et al. [17] (see Sect. 3.1.2).

They found that improvement of wear resistance for the AISI 420 sample plasma nitrided at 350 °C nitrided was much more than that of sample nitrided at 550 °C. The weight loss after 30 min wear test for untreated sample and the samples nitrided at 350 and 550 °C were 4.33 mg, 0.13 mg, and 0.21 mg respectively. This improved antiwear property was also observed by friction coefficients. As seen in

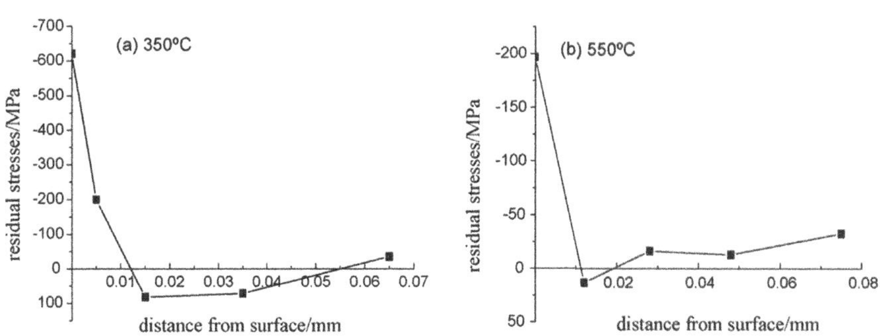

Fig. 3.31 Residual stress profiles of the layers nitrided at **a** 350 °C and **b** 550 °C [16]

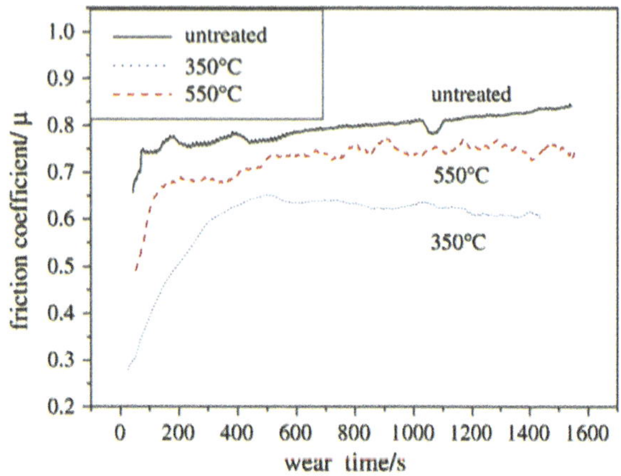

Fig. 3.32 Friction coefficient variations of the untreated and nitrided AISI 420 samples [17]

Fig. 3.32, in steady wear states, the friction coefficient values for the samples nitrided at 350 and 550 °C and unnitrided one were 0.6, 0.72, and about 0.8, respectively. As noted before, this improvement of the erosion resistance after plasma nitriding can be ascribed to several factors such as the microstructure, the high surface hardness (with resisting plastic deformation), and the high compressive residual stress (with closing the formed microcracks or impeding their formation during wear) in the nitrided layer.

Pinedo [19] (see Sect. 3.1.2) found a direct relation between hardness and nitriding duration for DIN X90 martensitic stainless steel. As seen in Fig. 3.33, the maximum hardness obtained with 12 h treatment ($\sim$1200 $HV_{0.1}$) was higher than that obtained for 8 h treatment ($\sim$1100 $HV_{0.1}$). This hardening effect can be attributed to the precipitation of chromium nitrides in the diffusion zone.

Espitia et al. [20] studied the cavitation erosion resistance of plasma nitrided of AISI 410 and martensitic stainless steel and compared to ASTM A743 grade CA6NM martensitic stainless steel.

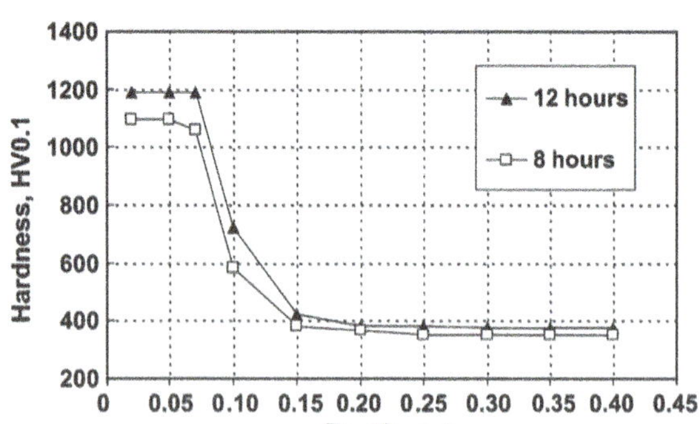

Fig. 3.33 Hardness variations with depth for DIN X90 sample plasma nitrided at 520 °C under atmosphere of N_2/H_2 = 1:1 [19]

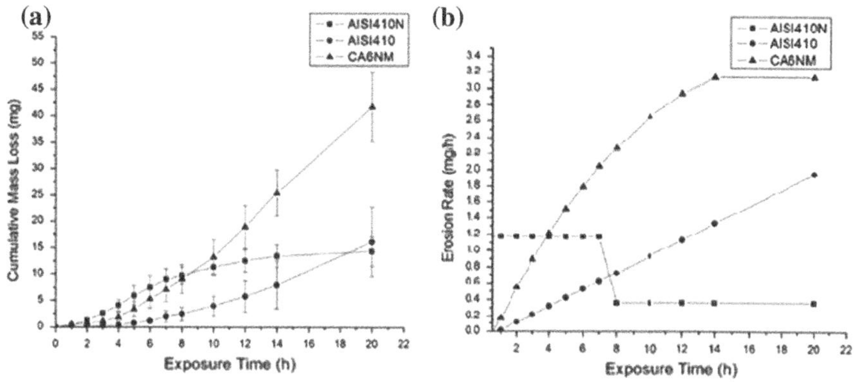

Fig. 3.34 Cumulative mass losses and erosion rates curves of samples, **a** cumulative mass loss and **b** erosion rate [20]

Figure 3.34 shows the variation of cumulative mass losses and the erosion rates with testing time, for all the samples. As can be seen, the nitrided AISI 410 showed the best and the CA6NM showed the worst behavior. The maximum erosion rates of nitrided AISI 410, CA6NM, and unnitrided AISI 410 steels were 1.2, 3.3, and 2 mg/h, respectively. As seen in Fig. 3.34, the cavitation rate of nitrided AISI 410 consisted of two steps with different erosion rates. The first stage is with an average erosion rate of 1.2 mg/h with duration of 7 h of testing and the second stage started from 8th hour of testing with an average erosion rate of 0.36 mg/h maintained till the end of testing. This drop was attributed to the removing of above-mentioned 5 μm thick nitrided layer and reaching of erosion to next layer of 20 μm thick.

CA6NM showed almost no mass loss before 2 h of testing. However, beyond 2 h, plastic deformation and increment of roughness were started. Finally after 20 h of testing, several grooves were observed and the fatigue and cleavage became predominant mechanism for wear. In the case of unnitrided AISI 410, the behavior was reported to be the same except in the rate of erosion which was less for this sample.

However, the situation was different for nitrided AISI 410. It was observed that the main mass loss mechanism in this sample was cracks coalescence under the surface and consequently detachment of large parts of the nitrided layer. The wear mechanisms observed on the nitrided AISI 410 sample surface were crack propagation and brittle detachment of grains without significant plastic deformation. Besides, it was observed that the surface roughness remained almost constant after 4th hour of testing indicating the removal of mass from a depth of almost 5 μm. After this depth (with starting the layer of just expanded martensite), the layer consisting of expanded martensite showed a better erosion behavior due to the constant roughness observed. It can be concluded that the presence of nitrides in nitrided layer was detrimental for erosion resistance.

Table 3.13 Roughness values R_a (μm) before and after plasma nitriding of the steels [21]

Nitriding T (°C)/*t* (h)	X6Cr17	X8CrNiTi18 10	X5CrNi18 10
–	0.012	0.17	0.02
300/60	0.18	0.23	0.06
350/26	0.23	0.21	0.11
350/40	0.40	0.20	0.16
500/20	0.64	0.87	0.61

3.2.3 Ferritic Stainless Steels

Larisch et al. [21] compared the surface roughness of some austenitic and ferritic stainless steels. It was found that surface roughness increased with temperature and time (Table 3.13). Plasma nitriding at low temperatures (<400 °C) and shorter durations led to a relatively low increase in surface roughness.

3.2.4 Tool Steels

Forati Rad et al. [22] (see Sect. 3.1.5) concluded that for AISI H11, the highest surface hardness was achieved for sample treated at 500 °C for 7 h. which contains the most percentage of ε phase. γ'-Fe_4N phase, has FCC crystalline structure and showed lower hardness with respect to ε-$Fe_{2\text{-}3}N$ phase with HCP crystalline structure. Also, because of transformation of ε to γ', the higher nitriding temperatures and longer nitriding durations, led to higher hardness. In the case of nitriding at 540 °C, the reduction of surface hardness could be attributed to further tempering of steel substrate.

The highest wear resistance belonged to the sample which was treated under 500 °C for 7 h that had the highest hardness and the greatest amount of ε phase. HCP crystalline structure had several active sliding systems in comparison with BCC and/or FCC. This led to showing better tribological properties in HCP structure.

The wear rate and wear weight loss of the samples plasma nitrided with 540 °C/7 h, 580 °C/5 h, and 580 °C/7 h conditions was higher than that of other nitrided samples. This was attributed to the thin compound layer formed on these samples and consequent breaking of this layer after small sliding distance which produces abrasive particles and results in the increasing of wear rate of the samples. However, plasma nitriding at 500 °C for 7 h improved the wear resistance more than other treatment conditions. As shown in Fig. 3.35, the compound layer formed on this sample resisted even after 2000 m sliding distance. The presence of a compound layer with appropriate thickness can improve the resistance of nitrided steel to both adhesive and abrasive wear.

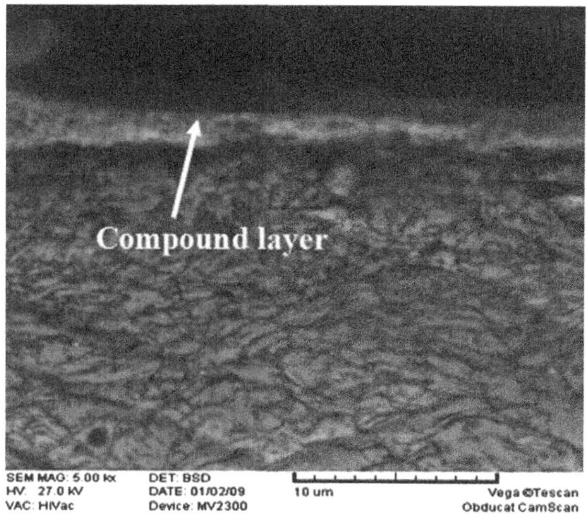

Fig. 3.35 Cross-sectional SEM micrograph of the sample plasma nitrided at 500 °C for 7 h after 2000 m sliding distance [22]

The improvement of wear properties for AISI H11 was also observed by Baki Karamiş et al. [23] (see Sect. 3.1.5). The untreated sample had the hardness of 790 $HV_{0.1}$, however for nitrided samples the hardness was 1200–1400 $HV_{0.1}$. Therefore, after wear test the weight loss for untreated sample was more severe than for nitrided samples.

They also concluded that the plasma nitriding temperatures or the hardness difference between the samples plasma nitrided at 500 and 550 °C did not severely affect the wear weight loss of nitrided samples.

In the case of AISI H13, Zagonel et al. [25] (see Sect. 3.1.5) observed that increasing process temperature caused hardness increment, i.e., increasing nitrogen incorporation. The hardness increase reached a depth of up to ~100 μm for all nitrided samples treated at temperatures higher than 330 °C. These results showed that at relative low temperatures applied in this study, an effective diffusion of nitrogen can occur.

For samples nitrided at 330 and 370 °C, the hardness gradient was observed. This was due to the accumulation of nitrogen on the top of the nitrided layer resulted by insufficient temperature for nitrogen diffusion. However, for samples nitrided at 420, 490, and 510 °C, a small hardness gradient was seen. The high hardness values obtained at the surface of nitrided samples (even at lower temperatures), suggested the existence of ε-$Fe_{2\text{-}3}N$ phase.

Soleimani et al. [26] (see Sect. 3.1.5) observed that using higher temperatures and longer durations led to the formation of harder surfaces for DIN 1.2210 cold work tool steel. The surface hardness was dependent on the nitriding conditions and varied in the range of 480–550 HV, whereas the core hardness did not change and remained constant at 336 HV. This may be explained as follows. At nitriding temperatures, a part of nitrogen atoms in the diffusion zone are dissolved in the ferrite and the remaining react with some alloying elements to form nitride

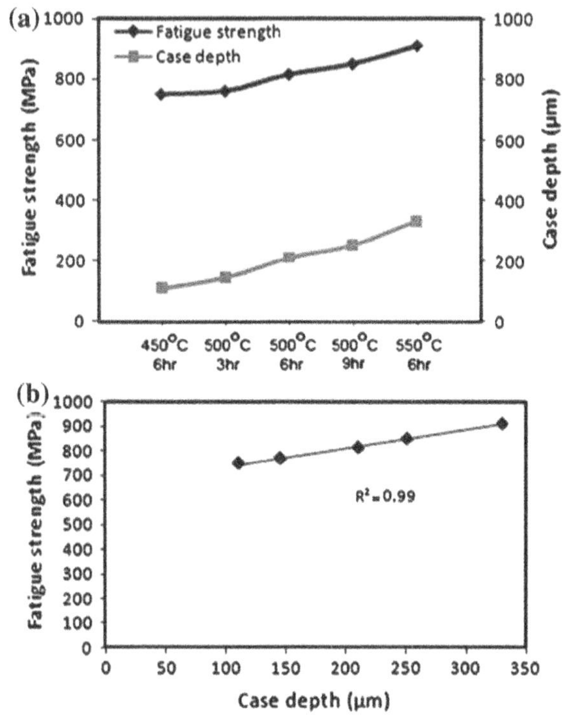

Fig. 3.36 **a** The variations of case depth and fatigue strength with plasma nitriding conditions and **b** the relationship between case depth and fatigue strength of plasma nitrided steel [26]

precipitates. Increasing nitriding temperature and time led to the fine dispersion of the nitride precipitates and consequently increase in the surface hardness.

In addition, increasing the nitriding temperature and time led to an increase in the case depth and consequently improvement of fatigue strength. Accordingly, a linear relationship was observed between the fatigue strength and case depth of the plasma nitrided sample which is shown in Fig. 3.36.

With increasing the surface hardness and case depth values, higher fatigue resistance was obtained. This could be due to the subsurface crack initiation instead of surface crack initiation that is normally seen in unnitrided samples. The subsurface fatigue cracks, called "fish eye" type, are initiated at nonmetallic inclusions near the case/core interface. Comparing the fatigue strength of the plasma nitrided samples at 550 °C for 6 h (910 MPa), which had the thickest case depth, with that of quenched and tempered specimen (545 MPa), an improvement of 67 % is seen. So the above discussion was proved.

Figure 3.37a shows the fracture surface of the sample plasma nitrided at 550 °C for 6 h in which typical subsurface fish eye crack can be seen. The sample was loaded under 950 MPa cyclic stress up to complete fracture at 627,000 cycles. As seen in Fig. 3.37a, these kinds of cracks are circular in shape which is ascribed to the pure rotating bending type of cyclic stressing.

Conci et al. [27] (see Sect. 3.1.5) investigated the effects of different nitriding gas mixtures on mechanical properties of the nitrided samples.

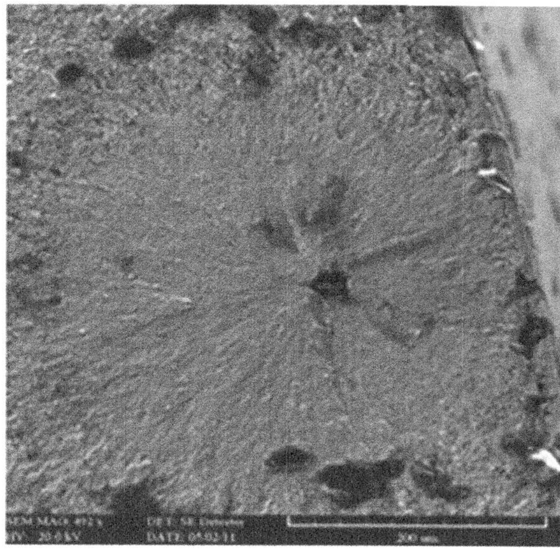

Fig. 3.37 The SEM micrograph of the fracture surface of the plasma nitrided fatigue test sample which shows a typical fish eye crack formation [26]

For nitrogen amounts of 5 and 10 % and short nitriding times, the maximum hardness achieved was approximately 1350 HV. However, further increase in treatment duration led to a drop to the range of 1180–1220 HV. This trend was also observed for atmosphere of 20 % N_2. For shorter durations, the surface microhardness was almost 1300 HV but by increasing the treatment time, this value dropped to about 1200 HV.

Figure 3.38 shows the wear volume of the samples treated under different conditions as well as the unnitrided sample. It is obvious that the lowest wear volume (0.13 mm^3) belonged to the sample nitrided in atmosphere of 20 % N_2 for 1 h. For nitrogen amounts of 10 and 5 %, the wear rate was 0.14 and 0.145 mm^3 respectively. As noted before, the coarsening and increasing the amount of the nitrides in the diffusion zone was resulted by increasing the nitrogen content in the atmosphere. This can lead to the loss in the fracture toughness of the surface consequently embrittlement of the surface and high tensile stress generated during the growth and coarsening of such nitrides. The formation of the brittle compound layer was another reason for deteriorating the wear resistance with increasing the both nitriding time and temperature.

3.2.5 Constructional and Structural Steels

Corengia et al. [29] (see Sect. 3.1.6) observed that with increasing nitriding time, the case/core interface got less conspicuous and the hardening effect affected a higher depth in AISI 4140 steel. Also the core hardness decreased with prolonging process as seen in Fig. 3.39. Additionally, the surface hardness increased with

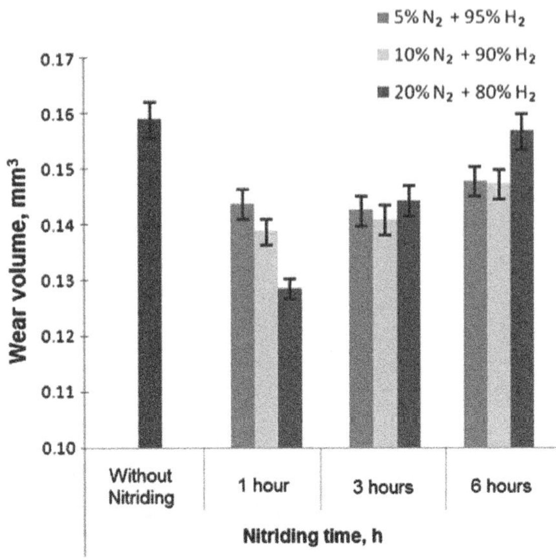

Fig. 3.38 Variations of the wear volume with nitriding time and nitrogen amount in the nitriding atmosphere for samples nitrided at 470 °C [27]

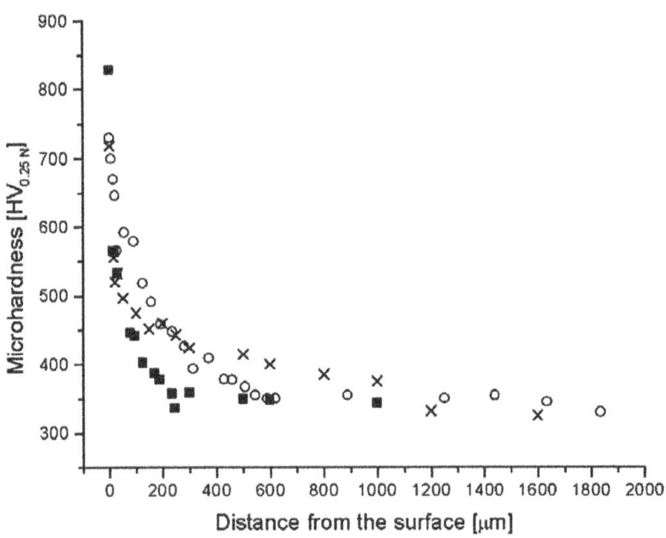

Fig. 3.39 Microhardness variations for pulsed DC plasma nitrided samples for black square 2 h, (o) 15 h, and (×) 28 h [29]

nitriding time and reached a maximum value at 2 h and then dropped to lower values at longer treatment durations. This indicated that the dispersion of nitrides precipitates in the matrix depended on the nitriding time.

It was concluded that the roughness increased with duration of plasma nitriding. During the first hours of nitriding, the rate of increase was higher which could be related to the relative intensity of sputtering and redeposition.

Ochoa et al. [30] (see Sect. 3.1.6) studied the effect of nitrogen content in different depths on the hardness. The data are plotted in Fig. 3.40. This plot shows that the behavior of hardness with nitrogen concentration was changed at around 20 at.% nitrogen concentration which indicated the structure changes of compound layer. Figure 3.18 shows the disappearance of the ε-$Fe_{2\text{-}3}N$ and γ'-Fe_4N nitride phases in this nitrogen concentration which caused the remaining nitrogen to be included in the α crystalline phase.

Li et al. [31] (see Sect. 3.1.6) observed that the maximum hardness value measured was around 930 $HV_{0.1}$ for the sample nitrided at 540 °C for 4 h in the case of 42CrMo low alloy steel. Temperature increment led to the deeper nitrogen diffusion into the substrate for both anodic and cathodic potential.

Podgornik et al. [4] (see Sect. 3.1.6) observed the roughness increment of plasma nitrided samples resulted by conically shaped asperities. With plasma nitriding in a 99.4 % H_2 + 0.6 % N_2 atmosphere for 17 h at 540 °C, the average roughness which was R_a = 0.35 μm for ground surface, increased to approximately 0.65 μm for AISI 4140 steel and 0.61 μm for AISI A355 steel samples. However, no difference was reported between surface roughness of the samples nitrided with conventional and pulsed DC plasma nitriding methods.

The average coefficients of friction for variously treated pins made from the test steels mated with a hardened ball bearing steel disc under different testing conditions showed that for both AISI 4140 and AISI A355 steels, the friction coefficient was mostly independent from plasma nitriding method (pulsed or conventional DC).

The friction coefficient remained almost constant during the dry sliding test method and was in the range of 0.5–0.65 for a low sliding speed (0.1 m/s) and 0.25-0.35 for a high sliding speed (1 m/s). This was an indication of changing contact condition with increasing sliding speed for both steels.

In the case of lubricated sliding test, the friction coefficient was reduced to about 0.1 for low load (60 N) and about 0.17 for 100 N load testing conditions, with no difference between the sliding speeds used. The enhanced friction coefficient when

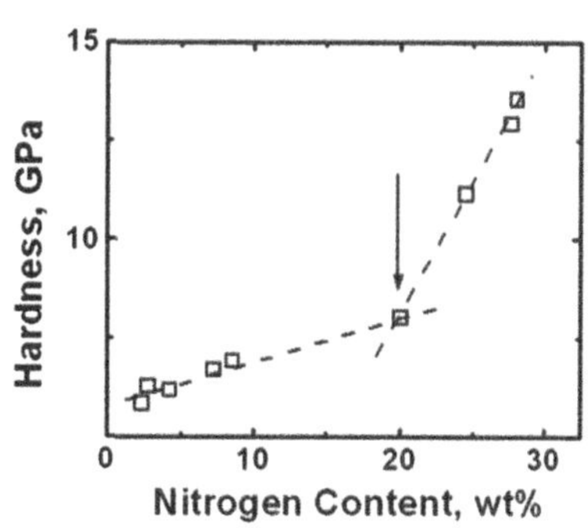

Fig. 3.40 Hardness variations with nitrogen concentration showing a structural change (see Fig. 3.18) [30]

applying higher load, can be ascribed to the steel–steel contacts which increased in number at higher load. In addition, lower friction coefficient values (up to 10 %) were obtained in the case of AISI A355 compared to the AISI 4140 steel. This behavior was shown because of higher surface hardness of AISI A355 which led to the better abrasive resistance of the nitrided surface.

Therefore, the improvement of wear resistance was observed after plasma nitriding for both studied steels under dry and lubricated sliding test.

3.3 Corrosion Properties

3.3.1 Austenitic Stainless Steel

Liang [5] (see Sect. 3.1.1) evaluated the corrosion resistance of plasma nitrided AISI 304 austenitic stainless steel. Potentiodynamic anodic polarization curves shown in Fig. 3.41 indicates that this steel had a very narrow passivation area in 3.5 % NaCl solution. After a narrow passive region, current again increased rapidly at about −10 mV. The sample plasma nitrided at 420 °C had a corrosion potential of −98 mV and exhibited a wide range of passivation as the potential increased. Higher temperature (465 °C) nitrided samples exhibited lower corrosion resistance in the 3.5 wt% NaCl solution which is indicated by its lower corrosion potential (−335 mV) and higher corrosion current density (1.6 μA cm^{-2}) compared to unnitrided sample (−215 mV, 68.8 nA cm^{-2}).

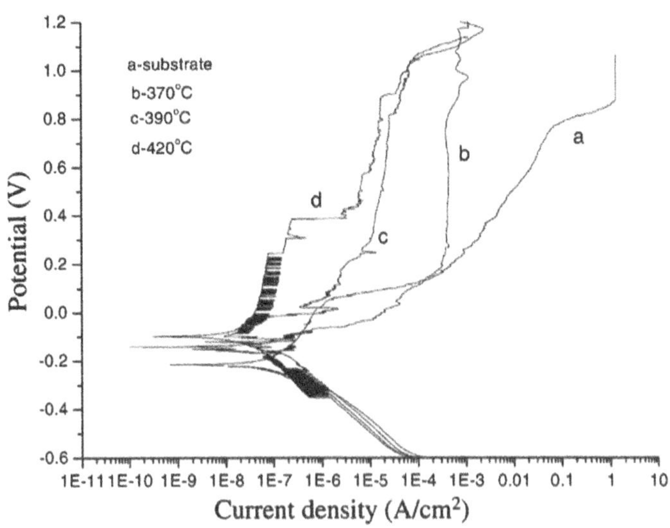

Fig. 3.41 Potentiodynamic curves of austenitic steels nitrided at lower temperatures in 3.5 % NaCl solution [5]

As can be seen in this figure, all of the graphs obtained from high temperature nitrided samples showed a sharp increase in the current density without a passive region. This means the breakdown of passivity in the passive oxide layer on the coating indicating localized corrosion, commonly known as pitting corrosion. In addition, the surface examination showed pits left on high temperature nitrided and unnitrided samples and no pits on low temperature nitrided specimens.

Gontijo et al. [8] (see Sect. 3.1.1) also confirmed the improvement of corrosion resistance at lower temperatures and its degradation at higher temperatures. It was reported that the surface became more reactive after treatment at higher temperatures compared to the treatment at lower temperatures.

The nitriding at lower temperatures increased the surface hardness without changing corrosion properties of the matrix. However, the existence of chromium compounds such as CrN, hardened the surface although, the surface became more reactive under anodic conditions and led to degradation of corrosion resistance at high temperatures.

The presence of deep and wide pits was reported for untreated samples, whereas for samples treated at lower temperatures, no corrosion pits were reported.

Menthe et al. [9] (see Sect. 3.1.1) observed improved uniform corrosion resistance for AISI 304L austenitic stainless steel measured in 5 N H_2SO_4 as obvious in Fig. 3.42. The initial critical current density was ten times higher than that of untreated sample. After removing parts of the surface layer step-by-step, the critical current density started to decrease and reached almost that of the untreated material.

Figure 3.43 shows behavior against pitting corrosion and high resistance against this type of corrosion. After removing the layer, corrosion resistance decreased but was still higher than untreated sample.

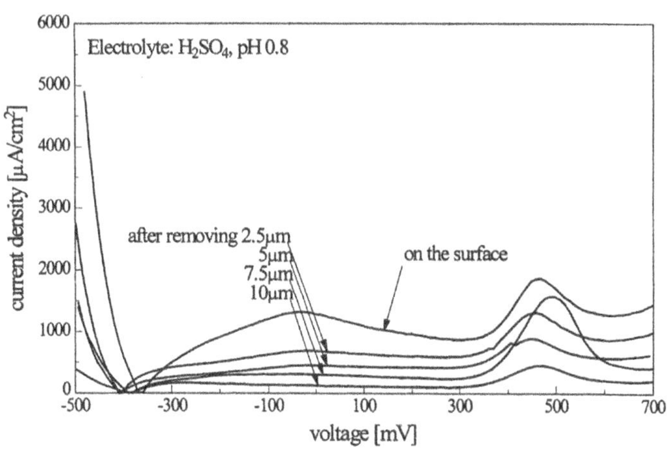

Fig. 3.42 Anodic polarization curves of plasma nitrided (5 h, 450 °C) samples measured after 10 cycles in 5 N H_2SO_4. The surface layer was removed step-by-step by grinding [9]

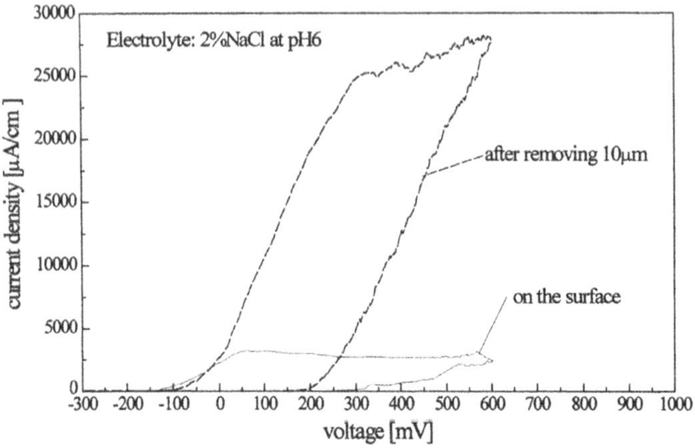

Fig. 3.43 Anodic polarization curves of plasma nitrided (5 h, 450 °C) samples measured after 10 cycles in 2 % NaCl solution. The surface layer was removed by grinding [9]

Corrosion tests results showed that the nitrided AISI 316L austenitic stainless steel exhibited almost the same corrosion resistance as the untreated 316L. The pitting corrosion was reported after corrosion tests [11].

Information about the corrosion behavior of the AISI 316L austenitic stainless steel which was evaluated by Li et al. [12] in NaCl solution is presented in Table 3.14. A passivation region was observed in the polarization curve of the unnitrided substrate. The current density was very low in the beginning, but rapidly increased at a potential of around 0.50 V. This was due to the breakdown of the passive film formed on the surface which led to the initiation of pitting corrosion in the presence of chloride ions.

For all the nitrided samples, corrosion resistance showed an improvement in comparison with unnitrided substrate which was indicated by higher corrosion potential of treated sample than that of unnitrided substrate (Table 3.14). For samples nitrided at 450 and 480 °C, the polarization curves showed a wide passive range with no pitting up to 1.2 V. At low potentials (−0.12 to 0.40 V), the current

Table 3.14 Corrosion resistance test results of the AISI 316L stainless steel in 3.5 wt% NaCl solution [12]

Samples	Corrosion potential E_{corr} (V)	Anodic current density J(A/cm^2), to E = 0.3 V	Pitting potential E_{pit} (V)	Anodic current density J(A/cm^2), to E = 1.2 V
Untreated	−0.258	2.50×10^{-6}	0.410	3.73×10^{-2}
450 °C	−0.252	7.19×10^{-6}	–	3.41×10^{-5}
480 °C	−0.193	3.04×10^{-5}	–	4.62×10^{-5}
510 °C	−0.223	1.43×10^{-5}	0.511	3.85×10^{-4}
540 °C	−0.156	3.10×10^{-6}	0.495	2.04×10^{-3}

density was little higher than that of the unnitrided sample. However, at higher potentials (0.50–1.2 V) the current density was significantly lower than that of the unnitrided sample. The passive behavior was also observed for high temperature (510 and 540 °C) nitrided samples. However, potential value for the beginning of pitting was only a little higher than that of the unnitrided substrate. In addition, the current density was lower than that of the unnitrided substrate above 0.70 V.

After completing corrosion test, some big and deep pits were observed on the surface of unnitrided sample. In contrast, no pitting was observed after test on the surface of the sample nitrided at 450 °C. With increasing the nitriding temperature to 480 °C, some small and shallow pits were observed in the corroded surface. Further temperature increment to 540 °C, led to the formation of more but shallower and smaller pits compared to untreated sample.

3.3.2 Martensitic Stainless Steel

Corengia et al. [14] reported the degradation of corrosion resistance of AISI 410 martensitic stainless steel after plasma nitriding. Figure 3.44 shows anodic polarization curves for an annealed steel sample in chloride media, in the presence of oxygen. As seen in this figure, anodic branch has a complex shape which is ascribed to the oscillation of the system between the active and passive states, resulting in the presence of multiple steady states. As can be seen, at more positive potentials, a well-defined passive region is observed up to approximately 0 V at which, the current increased suddenly due to pitting.

The sample tempered at 400 °C showed a well-defined passivation region after which, an increase of the current occurred. This behavior was indication of the sensitization effect. The occurrence of this phenomenon is along with the depletion of grain boundaries from chromium as a result of the formation of chromium carbides. It should be noted that, this effect was more significant at higher temperatures due to the nature of this phenomenon which is a thermally activated process. The sample tempered at 500 °C showed an active/passive behavior, followed by a poorly defined passivation area and an increase of the current at potentials positive to approximately −150 mV. The corrosion currents for the sample tempered at 500 °C were higher than those of sample tempered at 400 °C.

It can be seen in Fig. 3.45 that decreasing nitriding temperature resulted in a considerable shift in the corrosion potentials to more positive values. The samples nitrided at 400 and 500 °C experienced active dissolution at potentials positive to corrosion potential (E_{corr}). The sample nitrided at 350 °C showed the less negative corrosion potential and the lowest corrosion current density among the three samples tested.

With comparison of Figs. 3.44 and 3.45, it can be concluded that, the corrosion behavior was mainly determined by the depletion of chromium from the solid solution due to CrN precipitation. This effect largely predominated over the sensitization due to chromium carbides precipitation. Corrosion potential (E_{corr}) of the

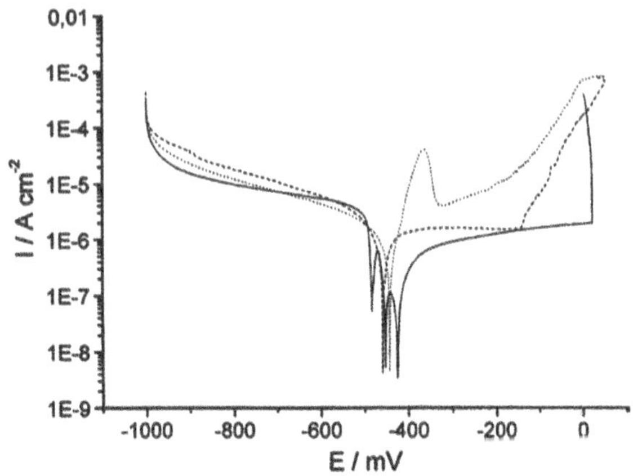

Fig. 3.44 Anodic polarization curves for: (—) unnitrided annealed substrate, (– –) annealed then tempered at 400 °C, (.....) annealed then tempered at 500 °C [14]

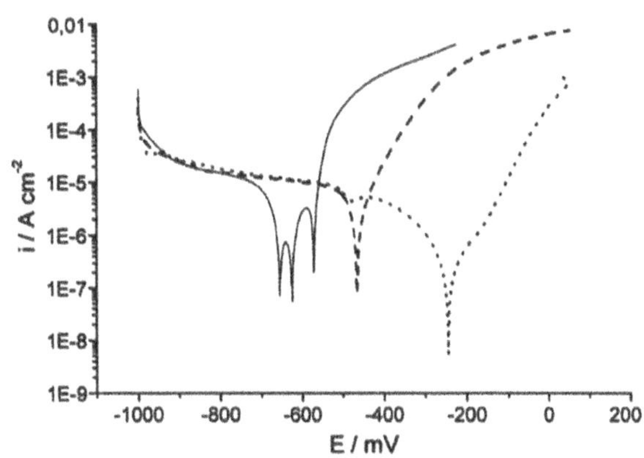

Fig. 3.45 Anodic polarization curves for annealed and then pulsed DC plasma nitrided samples for 20 h at (—) 500 °C, (– –) 400 °C and 350 °C (……) [14]

sample nitrided at 350 °C was more positive and its corrosion current (I_{corr}) was lower compared to the other samples studied which indicated a more noble behavior of the "expanded ferrite" phase in comparison with the chromium depleted solid solutions obtained at higher temperatures.

Therefore, it was concluded from this study that the microstructural aspects and corrosion properties of the annealed and quenched-tempered samples were similar.

Yun-tao et al. [16] (see Sect. 3.1.2) reported the decrease of corrosion resistance in the case of AISI 420 martensitic stainless steel plasma nitrided at higher temperatures.

Figure 3.46 shows the EIS test results of the nitrided and unnitrided samples after 30 min immersion in 5 wt% NaCl solution. As seen in this figure, a capacitive impedance arc is visible in the EIS plot of all the samples. The diameter of the capacitive impedance arc of the sample nitrided at 350 °C, showed an improvement compared to that of the unnitrided sample, while that of the sample nitrided at 550 °C was reduced by several times.

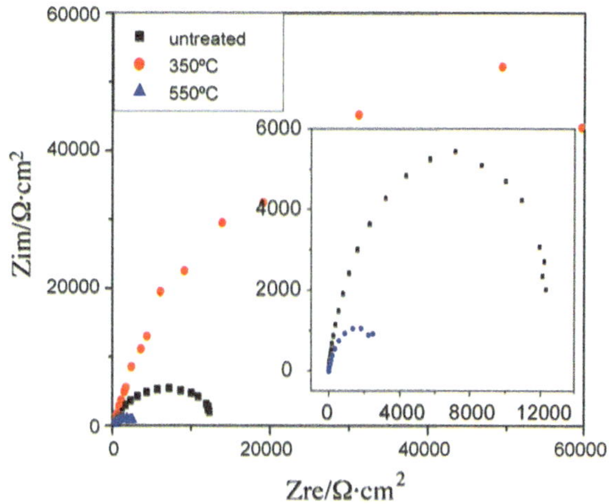

Fig. 3.46 Nyquist plot of samples in 5 % NaCl solution after 0.5 h immersion [16]

Nitriding at 550 °C deteriorated the corrosion resistance of AISI 420 steel. This was because of the CrN formation and the depletion of chromium in the solid solution of the nitrided layers. It should be noted that the presence of both Cr and N alloy elements is efficient for improving the resistance of the steel sample to localized corrosion. A dense and continuous oxide film was formed on the surface with combination of the chromium with oxygen and played a protective role to improve the corrosion resistance of the steel. During corrosion process, the dissolution of nitrogen and its reaction with hydrogen ions (H^+) occurs to form the $NH4^+$ and due to this cathodic reaction ($[N] + 4H^+ + 3e^- \rightarrow NH4^+$), the local pH increases which facilitates the re-passivation phenomenon. However, all of the mentioned steps completely happen when both nitrogen and chromium are in solid solution and their concentration is high enough. Therefore, with formation of CrN and the depletion of nitrogen and chromium in the solid solution of the nitrided layer, the passivation process was prevented and a reduction in the corrosion resistance was observed [17]. However, plasma nitriding at 350 °C, significantly improved the corrosion resistance of the AISI 420 steel. This was due to the prohibition of the CrN formation and the formation of ε-Fe_3N and α_N phases which have high chemical stability.

The erosion–corrosion behavior of samples was also studied in two environments. In both neutral and acid environment, nitriding at 350 °C was reported to improve the erosion–corrosion resistance of AISI 420 because it increased both the solid particle erosion resistance and corrosion resistance. In contrast, the erosion-corrosion resistance of the sample nitrided at 550 °C, was reported to be weaker than that of the unnitrided sample due to its lower capacity of corrosion resistance (Table 3.15).

Xi et al. [17] (see Sect. 3.1.2) did salt fog spray on the AISI 420 martensitic stainless steel and also observed the improvement of resistance to pitting corrosion in lower temperatures and vise versa.

Table 3.15 Results of the erosion–corrosion test of nitrided and unnitrided samples [16]

Experiment environment	Samples	Erosion–corrosion rate (g.m^{-2} h^{-1})	Comparisons
Neutral slurry (pH: 6.5–7)	Unnitrided	4.01	1
	350 °C nitrided	0.13	3.3 %
	550 °C nitrided	20.61	5.13 times
Acid slurry (pH: 3 - –0.2)	Unnitrided	23.66	1
	350 °C nitrided	2.98	12.6 %
	550 °C nitrided	35.17	1.49 times

The visual appearance of the unnitrided and nitrided samples after 120 h salt fog corrosion test is shown in Fig. 3.47. The red rust and numerous corrosion pits are obvious on the surface of the unnitrided sample as well as the samples nitrided at 450 and 550 °C. In addition, the highest corrosion damage belonged to the sample nitrided at 550 °C. However, for the sample nitrided at 350 °C, the improvement of corrosion resistance can be seen as shown in Fig. 3.47b.

Fig. 3.47 Visual aspects of the **a** untreated, **b** 350 °C, **c** 450 °C, and **d** 550 °C nitrided samples after 120 h salt fog corrosion test [17]

Brühl et al. [18] (see Sect. 3.1.2) did $CuSO_4$ test on the surface of M340 and N695 martensitic stainless steels. Before nitriding process, none of them showed sensitization after 6 min pent hydrated $CuSO_4$. However, for nitrided samples, copper deposition was observed in less than a minute. This showed the presence of free iron as an indication of incomplete passivation.

Results of the salt spray fog and electrochemical tests done by them also approved the diminishing of corrosion resistance after plasma nitriding.

In general, M340 and N695 martensitic stainless steels (both nitrided and unnitrided states) did not show an appropriate passive behavior which was indicated from high anodic currents observed within the tested potential range.

3.3.3 Precipitation Hardening Stainless Steels

Brühl et al. [18] (see Sect. 3.1.3) observed that the unnitrided Corrax precipitation hardening stainless steel, did not show sensitization, after 6 min of pent hydrated $CuSO_4$ and showed the same behavior. They also observed the preservation of corrosion resistance after salt spray test.

3.3.4 Tool Steels

Basso et al. [24] (see Sect. 3.1.5) evaluated corrosion resistance of AISI H13 tool steel. The OCP[1] values for nitrided samples in 0.9 % NaCl environment are shown in Table 3.16 (entries 1–6). It was reported that in the temperature range of 360–440 °C, the corrosion resistance increased due to the formation of a surface layer composed of ε-$Fe_{2\text{-}3}N$ which retained Cr atoms in its crystalline structure. Above 440 °C, corrosion resistance decreased due to the formation of CrN. With increasing the nitriding temperature to 520 °C, the resistance to pitting corrosion was significantly recovered as a result of the precipitation of both ε-$Fe_{2\text{-}3}N$ and γ'-Fe_4N nitride phases.

The mean OCP values measured after 30 min immersion (Table 3.16, entries 7–11) increased with nitriding time. However, a small decrease was observed for prolonged treatments because of the formation of γ'-Fe_4N nitride phase.

The improvement of resistance to pitting corrosion was also observed after plasma nitriding at relatively low temperature of 400 °C for different durations between 1 and 36 h. By nitriding at this temperature, the CrN precipitation was prohibited and the γ'-Fe_4N phase did not markedly influence the corrosion behavior, which remained almost unchanged for nitriding durations longer than 9 h.

---

[1]Open-circuit potential.

Table 3.16 Corrosion parameters obtained from cyclic potentiodynamic polarization tests in 0.9 % NaCl solutions for AISI H13 tool steel samples before and after plasma nitriding under different conditions [24]

Entry	Nitriding time	Substrate temperature (°C)	OCP (V/SCE)	E_{pit} (V/SCE)
1-H13	–	–	−0.797	−0.472
2	4	360	−0.529	−0.292
3	4	400	−0.490	+0.062
4	4	440	−0.265	+0.445
5	4	480	−0.555	−0.288
6	4	520	−0.288	+0.787
7	1	400	−0.520	−0.241
8	4	400	−0.490	+0.062
9	9	400	−0.357	+0.665
10	16	400	−0.395	+0.725
11	36	400	−0.424	+0.580

3.3.5 *Constructional and Structural Steels*

Li et al. [31] (see Sect. 3.1.6) observed the improvement of corrosion resistance in the case of plasma nitrided 42CrMo low alloy steel. Figure 3.48 presents the potentiodynamic curves of anodic polarization of samples nitrided for 4 h with both cathodic and anodic potentials as well as unnitrided 42CrMo steel in 3.5 % NaCl solution. A passivation behavior can be observed for all nitrided samples as indicated form shifting of the corrosion potential toward the more noble values in the potential range of −0.3 to 0.7 $V_{(SCE)}$. In contrast to nitrided samples, no passivation was observed for unnitrided sample. The comparison of the corrosion potential values and passivation ranges of nitrided samples with the unnitrided substrate indicates a significant improvement of corrosion resistance after plasma nitriding. The formation and development of a dense and thick compound layer with nitriding at 540 °C for 4 h corresponded for the high corrosion resistance of this sample. Formation of this layer assists passivation process and prohibits the penetration of Cl^- ions.

Figure 3.49 shows the surface morphologies of the unnitrided sample as well as samples nitrided at cathodic and anodic potentials after anodic polarization tests done up to potential of 1.2 $V_{(SCE)}$. As seen in Fig. 3.49a, unnitrided sample experienced a very severe corrosion which can be indicated from large amount of corrosion products and numerous deep pits on the surface after test. However, the pits on the corroded surface of nitrided samples were shallow and small in size and few in number for both cathodic and anodic potentials (Fig. 3.49b, c).

Basu et al. [32] (see Sect. 3.1.6) reported the improvement of corrosion resistance of SAE 52100 high carbon low alloy steel after plasma nitriding. It was reported that with increase in nitriding temperature, the corrosion potential (E_{corr}) increased, whereas the corrosion current density (I_{corr}) and corrosion rate decreased

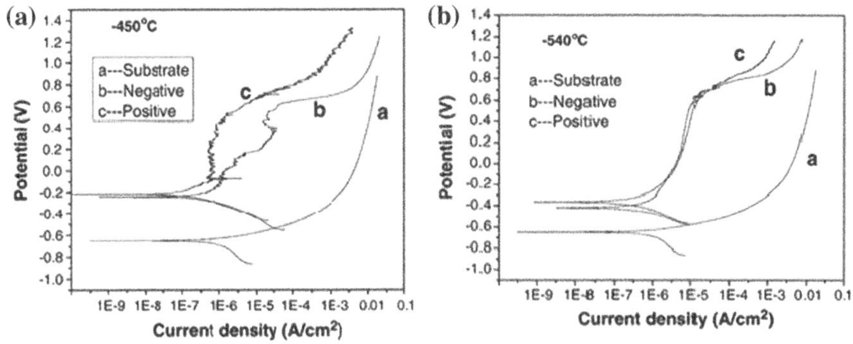

Fig. 3.48 Potentiodynamic curves of anodic polarization of unnitrided 42CrMo steel as well as samples nitrided at both anodic and cathodic potentials in 3.5 % NaCl solution [31]

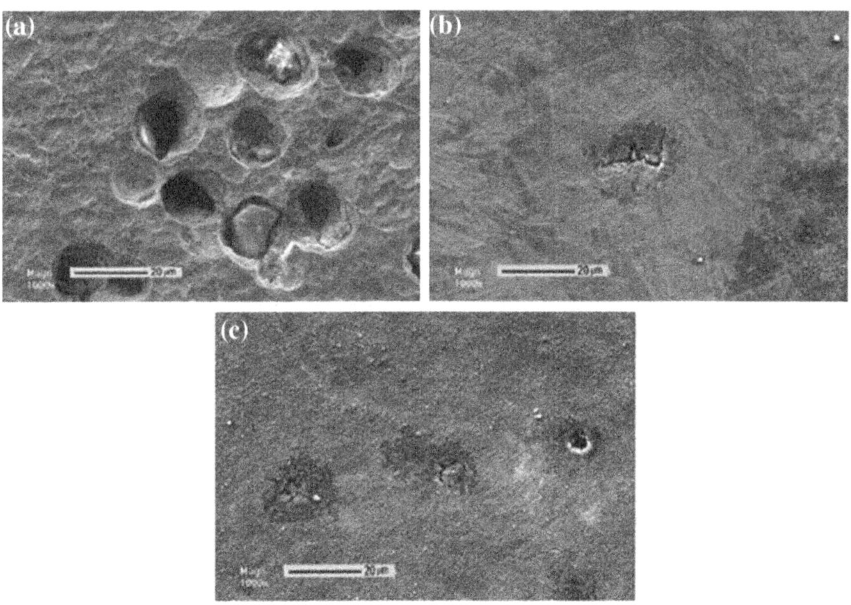

Fig. 3.49 Morphology of **a** untreated and nitrided 42CrMo steel at 540 °C for 4 h with **b** cathodic and **c** anodic potential after corrosion test [31]

and resistance to polarization (R_p) improved. With increasing nitriding time from 1 to 5 h, the same behavior was observed. The formation of a dense and compact nitride layer on the surface, corresponded for improvement of corrosion resistance. Due to its noble nature, formation of more nitride and thickening the layer, helps to cover the surface and protects it from corrosion attack. Therefore, an improvement was observed in corrosion resistance with increasing nitriding time and/or temperature.

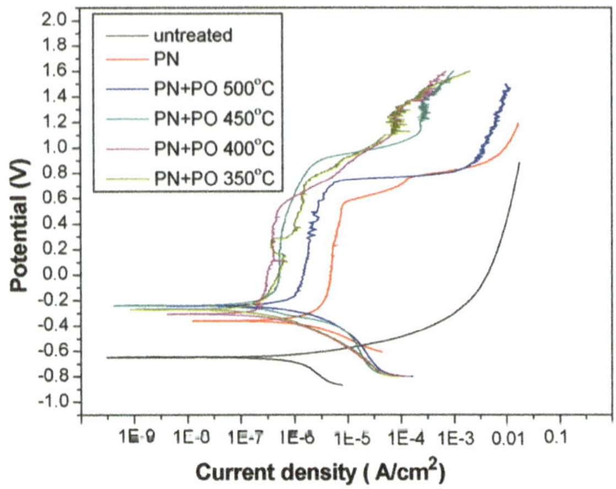

Fig. 3.50 Potentiodynamic polarization curves of untreated, plasma nitrided, and nitrided plus post-oxidized AISI 4140 steel samples [33]

Li et al. [33] (see Sect. 3.1.6) studied the effects of post-plasma oxidizing treatment on corrosion resistance of AISI 4140 steel.

Plasma nitriding in turn improved the corrosion resistance of samples because of the formation of a nitride layer composed of ε-$Fe_{2\text{-}3}N$. However, corrosion resistance was further improved with post-oxidation treatment as seen in Fig. 3.50. The values obtained for the corrosion potential were −0.27 $V_{(SCE)}$, −0.30 $V_{(SCE)}$, −0. 24 $V_{(SCE)}$, and −0.25 $V_{(SCE)}$ for the samples post-oxidized at temperatures of 350, 400, 450, and 500 °C, respectively. Despite the formation of relatively thick oxide layer of 2.5 μm at 500 °C, its corrosion resistance was low. This was owing to the fact that this temperature favored the formation of Fe_2O_3 phase and this phase is porous and loose. In contrast, the samples treated at 400 and 450 °C showed the best corrosion resistance due to the formation of mainly dense Fe_3O_4 phase.

Ebrahimi et al. [34] (see Sect. 3.1.6) investigated the effects of plasma oxidation after nitriding on corrosion resistance of AISI 4140.

For untreated and plasma nitrided samples, pitting corrosion was reported after corrosion test. But for post-oxidized samples no pitting was observed and just uniform corrosion was reported.

The electrochemical parameters obtained for all the samples are presented in Table 3.17. Among nitrided samples, the best corrosion resistance belonged to the sample nitrided at 570 °C as can be seen in Table 3.17. This was due to the higher amount of ε phase existed in the compound layer of this sample. However because of the thin layer formed on the sample nitrided at 530 °C, its corrosion resistance was lower than others.

Post-oxidation treatment further improved the corrosion resistance due to the covering and closing the surface pores by formation of oxide layer. According to Table 3.17, among the post-oxidized samples, the thickest oxide layer with thickness of about 3.9 μm was formed on the sample nitrided at 630 °C before

Table 3.17 Results of polarization test for the samples that were plasma nitrided before and after oxidation treatment [34]

Specimen	Corrosion rate (mpy)	E_{corr} (mV)	I_{corr} (μA)
Untreated	7.371	−655.6	12.53
Nitrided at 530 °C	7.195	−532.7	12.23
Nitrided at 570 °C	1.957	−519.2	3.325
Nitrided at 630 °C	3.473	−515	5.902
Nitrided at 530 °C-Oxidized	3.283	−370	5.578
Nitrided at 570 °C-Oxidized	1.344	−317.3	2.284
Nitrided at 630 °C-Oxidized	1.396	−366.1	2.373

oxidation. However, the corrosion resistance of this sample was lower than that of the sample nitrided at 570 °C before oxidation. This was probably owing to the fact that the surface roughness of the sample oxidized after nitriding at 630 °C was higher. Besides, the highest amount of Fe_3O_4 was formed in oxide layer of the sample nitrided at 570 °C and this significantly improved corrosion resistance due to its compact and adherent nature.

References

1. Podgornik B, Vizˇintin J, Leskovsˇek V (1998) Tribological properties of plasma and pulse plasma nitrided AISI 4140 steel. Surf Coat Technol 108–109:454–460
2. Jeong B Y, Kim M Ho (2001) Effects of the process parameters on the layer formation behavior of plasma nitrided steels. Surf Coat Technol 141:182–186
3. Alves Jr C, Rodrigues J A, Martinelli A E (1999) The effect of pulse width on the microstructure of d.c.-plasma-nitrided layers. Surf Coat Technol 122:112–117
4. Podgornik B, Vizˇintin J (2001) Wear resistance of pulse plasma nitrided AISI 4140 and A355 steels. Mater Sci Eng A315:28–34
5. Liang W (2003) Surface modification of AISI 304 austenitic stainless steel by plasma nitriding. Appl Surf Sci 211:308–314
6. Singh G P, Alphonsa J, Barhai P K, Rayjada P A, Raole P M, Mukherjee S (2006) Effect of surface roughness on the properties of the layer formed on AISI 304 stainless steel after plasma nitriding. Surf Coat Technol 200:5807–5811
7. Sharma M K, Saikia B K, Phukan A, Ganguli B (2006) Plasma nitriding of austenitic stainless steel in N_2 and N_2–H_2 dc pulsed discharge. Surf Coat Technol 201:2407–2413
8. Gontijo L C, Machado R, Kuri S E, Casteletti L C, Nascente P A P (2006) Corrosion resistance of the layers formed on the surface of plasma-nitrided AISI 304L steel. Thin Solid Films 515:1093–1096
9. Menthe E, Bulak A, Olfe J, Zimmermann A, Rie K T (2000) Improvement of the mechanical properties of austenitic stainless steel after plasma nitriding. Surf Coat Technol 133–1134:259–263
10. Jeong B Y, Kim M Ho (2001) Effects of pulse frequency and temperature on the nitride layer and surface characteristics of plasma nitrided stainless steel. Surf Coat Technol 137:249–254
11. Gil L, Brühl S, Jiménez L, Leon O, Guevara R, Staia M H (2006) Corrosion performance of the plasma nitrided 316L stainless steel. Surf Coat Technol 201:4424–4429

12. Li Y, Wang L, Xu J, Zhang D (2012) Plasma nitriding of AISI 316L austenitic stainless steels at anodic potential. Surf Coat Technol 206:2430–2437
13. Xia Y, Hu J, Zhou F, Lin Y, Qiao Y, Xu T (2005) Friction and wear behavior of plasma nitrided 1Cr18Ni9Ti austenitic stainless steel under lubrication condition. Mater Sci Eng A 402: 135–141
14. Corengia P, Ybarra G, Moina C, Cabo A, Broitman E (2004) Microstructure and corrosion behavior of DC-pulsed plasma nitrided AISI 410 martensitic stainless steel. Surf Coat Technol 187:63– 69
15. Corengia P, Walther F, Ybarra G, Sommadossi S, Corbari R, Broitman E (2006) Friction and rolling–sliding wear of DC-pulsed plasma nitrided AISI 410 martensitic stainless steel. Wear 260:479–485
16. Yun-tao X, Dao-xin L, Dong H (2008) Improvement of erosion and erosion–corrosion resistance of AISI420 stainless steel by low temperature plasma nitriding. Appl Surf Sci 254: 5953–5958
17. Yun-tao X, Dao-xin L, Dong H (2008) Improvement of corrosion and wear resistances of AISI 420 martensitic stainless steel using plasma nitriding at low temperature. Surf Coat Technol 202:2577–2583
18. Brühl S P, Charadia R, Simison S, Lamas D G, Cabo A (2010) Corrosion behavior of martensitic and precipitation hardening stainless steels treated by plasma nitriding. Surf Coat Technol 204:3280–3286
19. Pinedo C E (2003) The use of selective plasma nitriding on piston rings for performance improvement. Mater Des 24:131–135
20. Espitia L A, Varela L, Pinedo C E, Tschiptschin A P (2013) Cavitation erosion resistance of low temperature plasma nitrided martensitic stainless steel. Wear 301:449–456
21. Larisch B, Brusky U, Spies H J (1999) Plasma nitriding of stainless steels at low temperatures. Surf Coat Technol 116–119:205–211
22. Forati Rad H, Amadeh A, Moradi H (2011) Wear assessment of plasma nitrided AISI H11 steel. Mater Des, 32:2635–2643
23. Baki Karamıs M, Yıldızlı K, Aydın G C (2012) Sliding/rolling wear performance of plasma nitrided H11 hot working steel. Tribol Int 51:18–24
24. Basso R L O, Pastore H O, Schmidt V, Baumvol I J R, Abarca S A C, de Souza F S, Spinelli A, Figueroa C A, Giacomelli C (2010) Microstructure and corrosion behavior of pulsed plasma-nitrided ISI H13 tool steel. Corros Sci 52:3133–3139
25. Zagonel L F, Figueroa C A, Droppa Jr R, Alvarez F (2006) Influence of the process temperature on the steel microstructure and hardening in pulsed plasma nitriding. Surf Coat Technol 201:452–457
26. Soleimani S M Y, Mashreghi A R, Ghasemi S S, Moshrefifar M (2012) The effect of plasma nitriding on the fatigue behavior of DIN 1.2210 cold work tool steel. Mater Des 35:87–92
27. Conci M D, Bozzi A C, Franco Jr A R (2014) Effect of plasma nitriding potential on tribological behavior of AISI D2 cold-worked tool steel. Wear 317:188–193
28. Miola E J, de Souza S D, Olzon-Dionysio M, Spinelli D, dos Santos C A (1999) Nitriding of H-12 tool steel by direct-current and pulsed plasmas. Surf Coat Technol 116–119:347–351
29. Corengia P, Ybarra G, Moina C, Cabo A, Broitman E (2005) Microstructural and topographical studies of DC-pulsed plasma nitrided AISI 4140 low-alloy steel. Surf Coat Technol 200:2391– 2397
30. Ochoa E A, Wisnivesky D, Minea T, Ganciu M, Tauziede C, Chapon P, Alvarez F (2009) Microstructure and properties of the compound layer obtained by pulsed plasma nitriding in steel gears. Surf Coat Technol 203:1457–1461
31. Li Y, Wang L, Shen L, Zhang D, Wang C (2010) Plasma nitriding of 42CrMo low alloy steels at anodic or cathodic potentials. Surf Coat Technol 204:2337–2342
32. Basu A, Majumdar J D, Alphonsa J, Mukherjee S, Manna I (2008) Corrosion resistance improvement of high carbon low alloy steel by plasma nitriding. Mater Lett 62:3117–3120

33. Li Y, Wang L, Zhang D, Shen L (2010) Improvement of corrosion resistance of nitrided low alloy steel by plasma post-oxidation. Appl Surf Sci 256:4149–4152
34. Ebrahimi M, Heydarzadeh Sohi M, Honarbakhsh Raouf A, Mahboubi F (2010) Effect of plasma nitriding temperature on the corrosion behavior of AISI 4140 steel before and after oxidation. Surf Coat Technol 205:S261–S266

Chapter 4
Active Screen Plasma Nitriding

Conventional plasma nitriding has some limitations and disadvantages such as edge effect, hollow-cathode effect (The hollow-cathode effect usually seen in industry resulted by incorrect fixturing. This problem arises from insufficient distance between the parts to be nitrided. In this case, the glow discharge of near specimens is overlapped in cavities leads to the cascading of current densities and severe overheating of the parts [1, 2].) distortion and overheating of the specimen. Because of the above-mentioned problems, some new methods were presented to avoid the direct formation of plasma on the surface and consequent problems. One of the methods is **active screen** or **cathodic cage** plasma nitriding.

Active screen plasma nitriding technology is a novel nitriding process that the entire workload is surrounded by a metal screen or cage on which, a high voltage cathodic potential is applied. The formation of the plasma occurs on the screen rather than surface of the component to be treated. Therefore, ionic species do not interact directly with the surface. In this process, the sample is electrically isolated from the chamber. Under the high applied voltage, the sputtering of the cage walls and interaction with the nitriding atmosphere produce compounds and/or metallic species, which deposit on the sample's surface. Figure 4.1 presents a comparison between conventional and active screen plasma nitriding methods [3–7].

The main advantage of this method is related to the indirect sputtering (sputtering on the cathodic mesh, reaction of sputtered atoms with nitrogen and deposition of nitrides on the surface of the specimen). This leads to the reduction of arcing damage to the surface of the workpiece, elimination of edge and hollow-cathode effect and uniformity of thickness [6, 8, 9].

In this chapter, the effects of active screen plasma nitriding on microstructural, mechanical, and corrosion properties of steels will be discussed.

H. Aghajani and S. Behrangi, *Plasma Nitriding of Steels*, Topics in Mining, Metallurgy and Materials Engineering, DOI 10.1007/978-3-319-43068-3_4

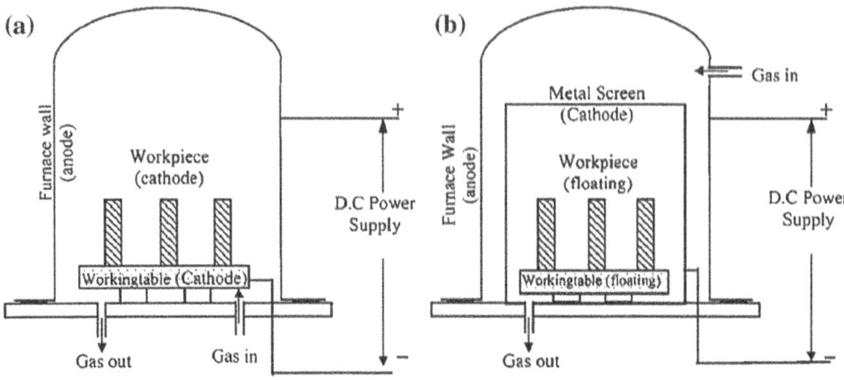

Fig. 4.1 Schematic of **a** conventional DC plasma nitriding and **b** active screen plasma nitriding [7]

4.1 Microstructure

4.1.1 Austenitic Stainless Steels

Gallo et al. [10] did active screen plasma nitriding on AISI 304 and 316 austenitic stainless steels and compared the results with conventional DC plasma nitrided samples. Two meshes were used, with holes 8 and 25.4 mm in diameter respectively. The sample-to-mesh distance was kept at 10 mm. Plasma nitriding was done at 420 °C, in a 25 % N_2–75 % H_2 atmosphere, under pressure of 400 Pa, for 10 h.

According to Fig. 4.2, the layer thickness achieved by active screen treatment was lower than conventional DC treatment.

The formation of expanded austenite was also reported for active screen treatment. As mentioned before, the formation of this phase caused the broadening and

Fig. 4.2 Variations of nitrogen content with depth for DC and active screen plasma nitrided AISI 316 samples [10]

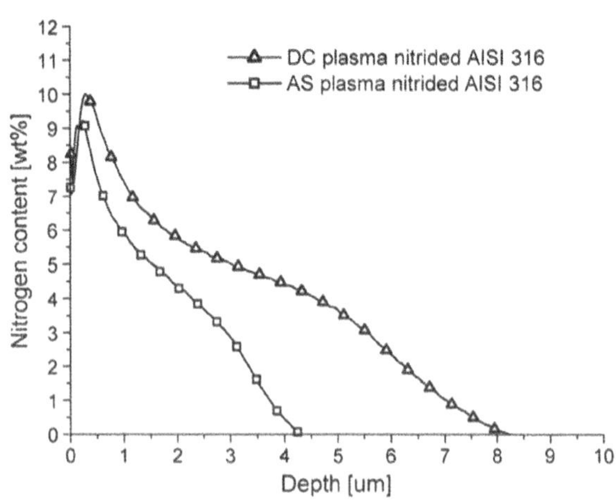

shifting of the XRD peaks to lower angles. However, the widening and shifting of the peaks were larger for the active screen treated samples than for the DC treated ones.

For a better suggestion about the mechanism of active screen plasma nitriding, a substrate of glass was used to identify the depositions formed during this process. The chemical composition of the deposited layer on the substrate indicated the sputtering of this material from the active screen mesh and consequent deposition on the treated samples. Some iron nitrides with different stoichiometric ratios were formed on the substrate.

Therefore, it was concluded that the most significant mechanism for active screen plasma nitriding is the sputtering and redeposition mechanism. Furthermore, transfer of material from the mesh to the worktable took place (Fig. 4.3).

de Sousa et al. [11] investigated the influence of sample position (Fig. 4.4) in active screen plasma nitriding of AISI 316 austenitic stainless steel. The cathodic cage they used was made of AISI 316 with 0.8 mm thickness, 112 mm diameter, 25 mm height, 8 mm width and uniformly distributed holes with a distance of 9.2 mm between centers of adjacent holes. Plasma nitriding was done at 450 °C, pressures of 120 and 250 Pa, in 80 % N_2 + 20 % H_2 atmosphere for 5 h.

According to Table 4.1, it can be concluded that the lower the pressure, the thicker the nitrided layer. Also, the samples that placed nearer to the screen showed a slight increase in the layer thickness. It can be explained by the superposition of the species flow from the top and lateral screen walls and consequently species flow rate. With decrease in the pressure, the mean free path for the ion particles that reached the holes of the cathodic cage walls decreased, increasing the sputtering rate. This led to increasing of the particle flow, and obtaining higher deposition efficiency. Therefore, thicker layers could be obtained compared to conventional DC plasma nitriding. It was concluded that, in the cathodic cage technique, deposition is the predominant process.

In the pressure of 250 Pa, the sample placed in position C, was reported to show the presence of expanded austenite. For position A, the high energy and intensity of

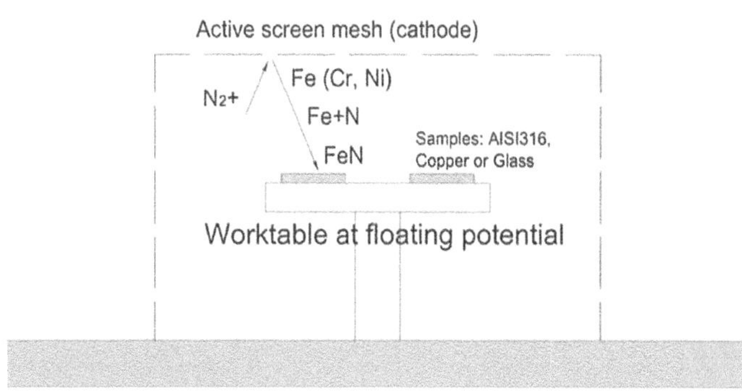

Fig. 4.3 Schematic of the sputtering and redeposition mechanism suggested for active screen plasma nitriding [10]

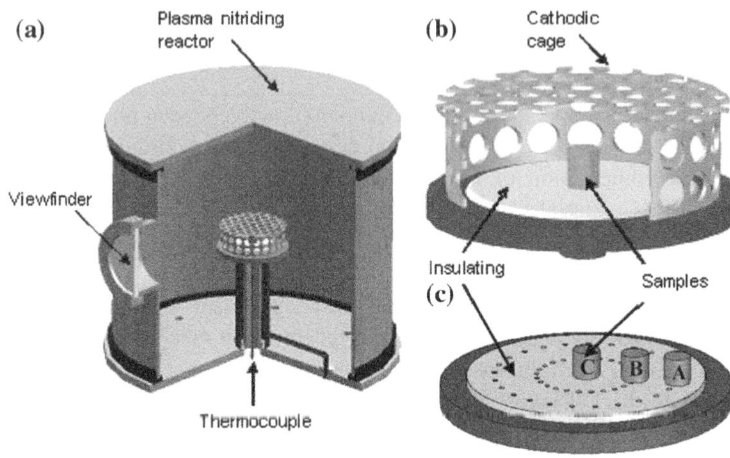

Fig. 4.4 **a** The reactor scheme of plasma ionic nitride, **b** configuration of cathodic cage used onto process and **c** spatial distribution of the samples nitrided at 250 Pa [11]

Table 4.1 Thicknesses and microhardness values of the nitrided layers as a function of the treatment pressure and sample position inside the cage (nitriding conditions: 5 h, 350 °C and 80 % N_2 + 20 % H_2) [11]

Pressure (Pa)	Sample position	Nitriding layer thickness (μm)	Microhardness ($HV_{0.1}$) ± 8.0 %
250	A	28.2 ± 0.7	1115
	B	24.4 ± 0.8	1305
	C	22.2 ± 1.0	1213
120	A	61.9 ± 1.0	978
	B	56.7 ± 1.5	953
	C	51.4 ± 1.6	931

the species reached the surface of sample and iron nitrides became predominant on the surface.

However, for samples nitrided under pressure of 120 Pa, the expanded austenite phase was not detected.

Alves Jr. et al. [12] plasma nitrided AISI 316 austenitic stainless steel samples at 500 °C, for 5 h, under pressure of 360 Pa and in 80 % N_2–H_2 atmosphere using active screen technique. Besides, DC plasma nitriding was done with applying same parameters except the atmosphere that they have selected 20 % N_2–H_2 for this treatment. The screen had these properties: diameter of 76 mm, height of 25 mm, thickness of 0.8 mm, diameter of holes of 7.6 mm and the holes' distance of 9.2 mm.

It was observed that, in active screen plasma nitrided samples, the surface was uniform and dark gray in color. However in DC plasma nitrided ones, the edge

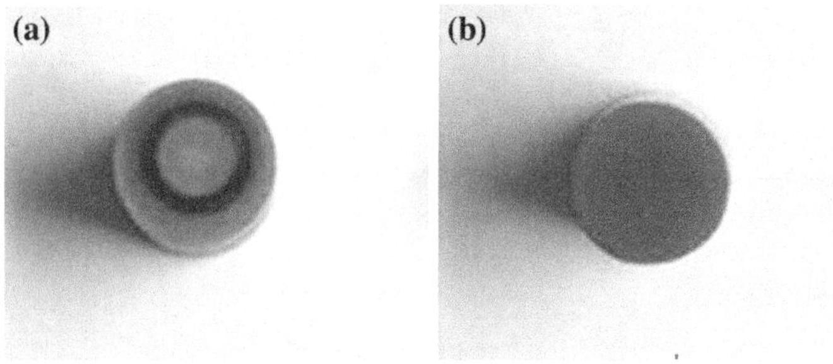

Fig. 4.5 Visual aspect of steel samples treated by **a** DC and **b** active screen plasma nitriding [12]

effect was clear as obvious in Fig. 4.5 (This effect was also reported by other authors like Li et al. [13]). This was corresponded to the fact that, in active screen treatment, plasma is not formed directly on the sample so the heat radiation comes from screen.

In case of thickness and morphology, no clear difference was observed between the samples nitrided using two methods at the same temperature.

It was reported that the formation of γ'-Fe_4N was higher for DC plasma nitrided samples due to direct interaction of the plasma species with the sample surface, which resulted in a larger rate sputtering and decarburizing. However, using active screen led to the elimination of surface decarburizing and consequently the larger formation of ε-$Fe_{2\text{-}3}N$. This was owing to the fact that, ε phase needs more nitrogen to be formed and the amount of nitrogen in the surface of active screen plasma nitrided samples was higher than DC plasma nitrided ones.

de Sousa et al. [14] investigated the effect of different gas mixtures on the active screen plasma nitrided AISI 316 austenitic stainless steel samples. The nitriding was done at 450 °C, for 5 h, under pressure of 250 Pa. The treatment was done using different gas mixtures: 20 % N_2 + 80 % H_2, 50 % N_2 + 50 % H_2, 80 % N_2 + 20 % H_2, and 100 % N_2. The cathodic cage was 112 mm in diameter, 25 mm in height and had 8 mm diameter holes. The holes were uniformly distributed at 9.2 mm on center.

Figure 4.6 shows cross-sectional micrographs of stainless steel AISI 316 samples nitrided under different atmospheres. The thickest layer was observed for samples treated with atmosphere of 80 % N_2 and 100 % N_2. As seen in this figure, the nitrided layer thickened with increasing the nitrogen content in the atmosphere up to 80 %, but the thickness decreased when using 100 % nitrogen atmosphere.

It was reported that for all samples, the Fe_xN nitrides were present for all of gas compositions. However, lower nitrogen concentration (20 %) led to the formation of CrN. For higher nitrogen contents, nitride phases such as FeN, Fe_3N, and Fe_4N appeared. The formation of expanded austenite was also observed for samples nitrided under atmospheres with nitrogen content ranging from 80 to 100 %.

Li et al. [15] also reported the formation of expanded austenite at lower and CrN at higher temperatures.

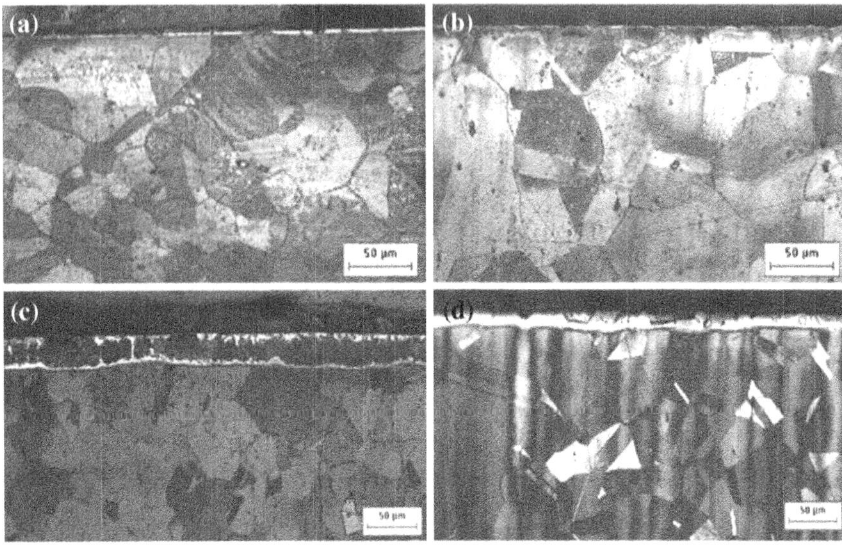

Fig. 4.6 Cross-sectional micrographs of the samples plasma nitrided in different atmospheres: **a** 20 % N_2, **b** 50 % N_2, **c** 80 % N_2 and **d** 100 % N_2 [14]

Li et al. [16] studied active screen plasma nitriding of AISI 316 austenitic stainless steel. Plasma nitriding was done at 420 and 500 °C, under pressure of 500 Pa, for 20 h in an atmosphere of 25 % N_2 + 75 % H_2.

It was reported that at 420 °C, the nitrided layer had the thickness of 10 µm. The amount of nitrogen in the surface of the nitrided layer was very high and slowly reduced with increasing distance from surface until the substrate nitrogen value at a depth of about 12 µm. Besides, expanded austenite was formed at this temperature.

In contrast, nitriding at 500 °C produced a thicker nitrided layer. In this case, the presence of a bright sublayer was observed between the dark nitrided layer and the substrate as seen in Fig. 4.7a. This bright layer was characterized by the precipitation of chromium nitride. A very thin layer of 1 µm thick was also observed on top of the nitrided surface as shown in Fig. 4.7b. It was reported that, this layer can just belong to active screen plasma nitriding and constituted by γ'-Fe_4N and CrN due to the deposition conditions of active screen method.

de Sousa et al. [17] evaluated the effects of adding CH_4 (methane) gas to the atmosphere of plasma nitriding setup. They treated the samples at temperatures of 400, 450, and 500 °C, under pressure of 250 and 500 Pa, for 5 h and in atmosphere of 95 % N_2 + 5 % CH_4. The active screen was made from AISI 316 austenitic stainless steel with thickness of 0.8 mm, diameter of 112 mm, and height of 25 mm with 8 mm diameter holes uniformly distributed with a distance of 9.2 mm between centers of adjacent holes.

The formation of a compound dual-layer was reported for all of conditions. It was observed that the inner layer consisted mainly of carbon and the outer one

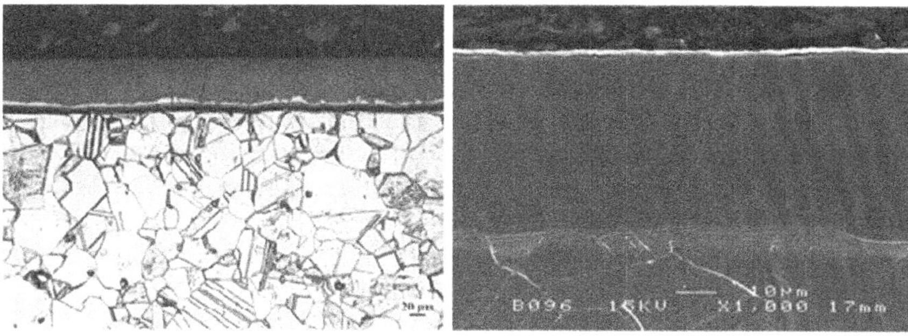

(a) Optical microstructure (b) SEM

Fig. 4.7 Microstructure of the active screen plasma nitrided AISI 316 sample (etched in 50 % HCl–25 % HNO_3–25 % H_2O for 1 min) [16]

consists of nitrogen. The thicknesses of the nitrided layers varied very slightly as a function of pressure at a constant temperature. The thickening of nitrided layers was observed with an increment in nitriding temperature. At higher temperatures, the thickness reached to values of up to 30 μm.

The phase composition of samples plasma nitrided with addition of small amounts of methane had some differences compared to the samples nitrided without presence of methane gas. Under pressure of 500 Pa and at higher temperatures of 450 and 500 °C, some traces of CrN phase appeared due to an insignificant precipitation which led to a higher corrosion resistance. At this pressure, expanded austenite and ε-Fe_3N were also present in the nitrided layer. Under lower pressure of 250 Pa, only expanded austenite as well as ε-Fe_3N was present without any traces of chromium nitride.

Nishimoto et al. [18] investigated effect of the distance between screen and sample on active screen plasma nitriding properties of low carbon S15C steel and SUS 304 austenitic stainless steel. The screen used was made of SUS 304 with a diameter of 150 mm and a height of 155 mm with the thickness of 0.5 mm. The distance between sample and screen was adjusted to be 10, 20, 30, 40, and 50 mm. Plasma nitriding was done at 450 °C in atmosphere of 25 % N_2 + 75 % H_2 for 18 ks under pressure of 600 Pa.

It was reported that for none of the samples the edge effect was observed due to the indirect heating of the sample and homogenous increment of temperature in the sample.

It was reported that for both steels the phase constitution remained constant regardless of the distance between the sample and screen. The formed phases on the surface of the S15C sample were ε-$Fe_{2\text{-}3}N$, γ′-Fe_4N, and α-ferrite. However γ-austenite and expanded austenite were the phases formed on the surface of the SUS 304 sample.

It was reported that the morphology of the particles formed on the surface was polygonal and spherical when the distance between the screen and the sample was 10 and 50 mm, respectively. The refinement of the particle size was also observed with increasing distance between the screen and the sample for both steels. This could be owing to the presence of Fe_xN particles in the plasma that were formed on the active screen and deposited on the sample surface during active screen plasma nitriding. This was clear that decreasing the distance between screen and sample led to the growth of these particles.

The effect of increasing distance was also observed on the thickness of the compound layer for S15C and expanded austenite layer for SU304.

4.1.2 Structural and Constructional Steels

Ahangarani et al. [19] compared the plasma nitriding response of 30CrNiMo8 low alloy steel treated with two methods: conventional DC and active screen plasma nitriding under different conditions. The active screen was a mesh cylinder of 70 mm diameter and 50 mm height, made of 0.8 mm thickness perforated st37 steel sheet, which equally distributed around holes of 6 mm (and 8 mm) diameter, with a detachable top lid. The top lid was made of the same mesh sheet as cylinder (and of st37 steel plate of 0.8 mm thick). The distance between the sample surface and the top lid was 12 mm. The nearest distance from the cylinder to the sample edge was about 10 mm.

As shown in Table 4.2, compound layer thickness of DC plasma nitrided and active screen plasma nitrided samples depended on temperature, treatment time, and screen setup parameters. For active screen plasma nitriding, there was a direct relation between the temperature and time with the thicknesses of compound layer and diffusion zone (for both screen and iron plate lid and both gas mixtures). For DC plasma nitriding, in 5 h, the thickness increased with temperature but in 10 h decreased due to sputtering rate increment on the sample surface. In both

Table 4.2 Variations of compound layer thickness with temperature and screen setup parameters for both active screen plasma nitriding and DC plasma nitriding at 550 and 580 °C for 5 h [19]

Temperature (°C)	Compound layer (μm) for DCPN				Screen setup parameters		Compound layer (μm) for $AS_{8\ mm}$ PN			
	75 N_2/25 H_2		25 N_2/75 H_2		Top lid	Screen hole size (mm)	75 N_2/25 H_2		25 N_2/75 H_2	
	5 h	10 h	5 h	10 h			5 h	10 h	5 h	10 h
550	6	15.5	4	13	Iron plate	8	12	12.8	2	8.5
					Screen (8 mm)	8	10	13.6	5	10
580	12	13.5	11	11	Iron plate	8	12.6	13.2	10.5	12
					Screen (8 mm)	8	14.4	15.1	12	12.8

techniques, the thickness increased with an increase in the amount of nitrogen in the atmosphere due to enhancement of nitrogen atoms diffusion.

Figure 4.8 shows SEM cross sections of samples which were active screen plasma nitrided for 10 h at 550 and 580 °C with screen top lid in both 75 % N_2 + 25 H_2 (a and c) and 25 % N_2 + 75 H_2 (b and d) atmospheres. It is obvious in this figure that the layer thickened amount of N_2 in the nitriding atmosphere and temperature. Table 4.2 shows that use of iron plate top lid for nitriding at 550 °C could be more effective to decrease the thickness of treated layer.

It was reported that both γ′-Fe_4N and ε-$Fe_{2\text{-}3}N$ phases were formed. In the active screen plasma nitriding process, while the nitriding time increased, the amount of ε phase in the compound layer increased for all screen setups. When screen top lid was used, the amount of ε phase was higher than that of iron plate top lid. In the DC plasma nitriding process, the amount of ε phase in the compound layer slightly varied for 75 % N_2 + 25 % H_2 atmosphere but decreased for 25 % N_2 + 75 % H_2 atmosphere regardless of the treatment temperature.

Treating by both processes led to the formation of layer composed of γ′-Fe_4N and ε-$Fe_{2-3}N$ phases. However, the relative amounts of the two phases were different in samples treated with different conditions indicating the effectiveness of screen setup parameters in assisting the transition of active particles toward inside the screen setup and around samples. Furthermore, temperature increment directly increased sputtering rate on the screen tube and had no effect on the sample surface.

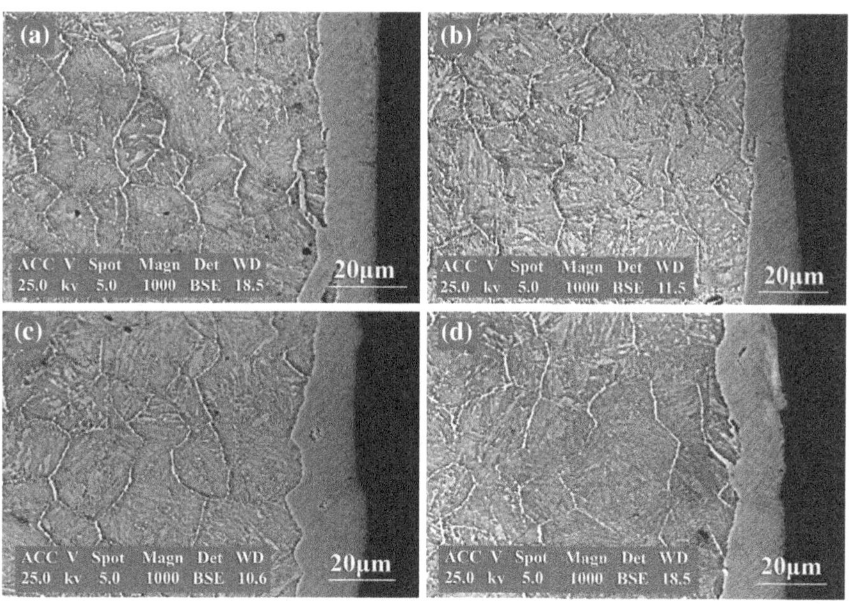

Fig. 4.8 SEM cross-sectional micrographs of 30CrNiMo8 steel active screen plasma nitrided at **a**, **b** 550 °C and **c**, **d** 580 °C for 10 h, screen top lid, at **a**, **c** 75 % N_2 + 25 H_2 and **b**, **d** 25 % N_2 + 75 H_2 atmosphere [19]

This showed that ion transition via sputtering and re-condensation mechanism of screen top lid on the sample surface is more effective on the layer thickness than iron plate top lid.

For a fixed hole size of screen setup (8 mm) at atmosphere of 75 % N_2 + 25 % H_2, for both iron plate and screen top lid, prolonging the treatment had little effect on the amount of ε phase in the compound layer. The reason was that prolonging the treatment directly increased sputtering rate on the screen tube and had no effect on the sample surface.

For DC plasma nitriding, it was reported that for both gas mixtures and both nitriding times used in this study, the amount of ε phase in the compound layer decreased with increasing treatment time and temperature. Against, amount of ε phase for both nitriding temperatures (550 and 580 °C) increased with nitrogen amount in the nitriding atmosphere. It was observed that using 75 % N_2 + 25 % H_2 gas mixture as nitriding atmosphere, led to the formation of sufficient ε amount regardless of the treatment temperature.

Figure 4.9 shows SEM images of 30CrNiMo8 low alloy steel DC plasma nitrided (a) and active screen plasma nitrided (b) at 550 °C for 5 h. Figure 4.9b and d (with higher magnification) shows that compound layer surface of active screen plasma nitrided samples consisted of hexagonal particles (constituting mainly of ε-phase) regularly distributed with well-defined boundaries. However, the particles on the sample surface in Fig. 4.9a, c (with higher magnification) have irregular size and distribution, probably, due to the continuous sputtering on the cathodic sample surface during nitriding.

The above-mentioned results were confirmed in the authors' other article [20].

Ahangarani et al. [21] investigated the effects of screen hole size on the plasma nitriding response of the 30CrNiMo8 low alloy steel. All of the conditions were the same as the previous articles of these authors. However, the diameter of screen holes was changed.

It was observed that for screen top lid conditions, the compound layer and diffusion zone thickened with increasing treatment temperature, but increasing of screen hole size at the same temperature had no considerable effect on the layer

Fig. 4.9 SEM micrographs of 30CrNiMo8 low alloy steel **a**, **c** DC plasma nitrided and **b**, **d** active screen plasma nitrided (8 mm) with screen top lid at 550 °C for 5 h [19]

Fig. 4.10 Optical microscopy images of 30CrNiMo8 low-alloy steel active screen plasma nitrided at 550 °C for 5 h, with iron plate top lid and holes' diameters of **a** 6 mm and **b** 8 mm [21]

thickness. However, for iron plate conditions, thickness of the layer increased with an increase in screen hole size at the same temperature (Fig. 4.10).

Nitrided layers produced with active screen plasma nitriding at 520, 550, and 580 °C for 5 h were different in terms of morphology, thickness, and phase structure. Because of the smallest transition of active species through the top lid towards the sample's surface, the least thick compound layer was formed at the screen setups with 6 mm hole size and iron plate top lid at 520 °C for 5 h. With increasing the hole size of the screen setup, the amount of the ε phase in the compound layer increased for both screen and iron plate top lid.

Alves Jr. et al. [12] studied active screen plasma nitriding of AISI 1020 as well as AISI 316 and observed the same results. The so-called edge effect was also observed in these samples as seen in Fig. 4.11.

de Sousa et al. [22] investigated active screen plasma nitriding of AISI 1020. Treatments were done at 400, 500 and 670 °C, for 3 h, under pressure of 360 Pa and gas mixture of 80 % N_2 + H_2. Besides, samples with different heights of 1, 3, 5, 8, and 10 mm were prepared. The cathodic cage was 0.8 mm thick AISI 316

Fig. 4.11 Visual aspects of samples nitrided by **a** DC plasma nitriding and **b** active screen plasma nitriding [12]

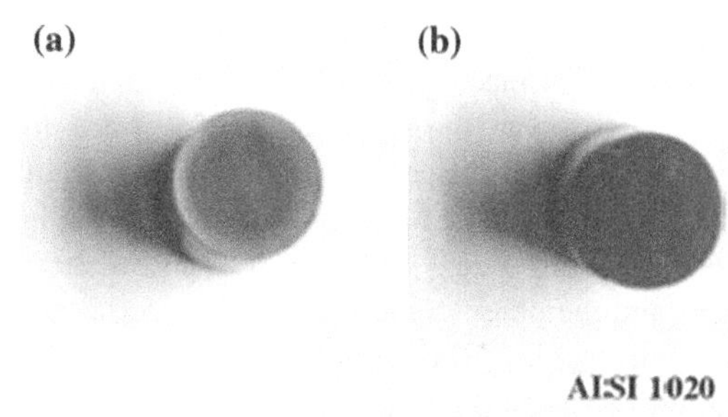

austenitic stainless steel, 112 mm in diameter and 25 mm in height, with holes of 8 mm in diameter spread uniformly, with 9.2 mm between centers of adjacent holes.

It was observed that, for different sample height, the compound layer thicknesses were the same for each parameter. Also no considerable change of the thickness was reported for different nitriding temperatures.

The XRD patterns presented in Fig. 4.5 obviously show that the same phases were present in all samples regardless of the sample height and temperature. There was also a peak belongs to α-Fe as well as Cr. The presence of Cr was attributed to the material (stainless steel) from which the cathodic cage was made. No significant variation in the intensities of γ' and ε phases is seen in Fig. 4.12.

Li et al. [23] studied active screen plasma nitriding of 722M24 low alloy steel. Samples were treated at 500 °C, for 5 h, under pressure of 500 Pa and three atmospheres: 25 % N_2 + 75 % H_2, 25 % N_2 + 75 % Ar and 100 % N_2. The screen used was of 120 mm diameter and 130 mm height. The cylinder was made of 0.7 mm thick perforated steel sheet, which had uniformly distributed round holes of 8 mm diameter. The center to center distance between each neighboring hole was 9.2 mm, accordingly, the open area fraction of the mesh was 68 %. It had a removable top lid made of low carbon steel plate with 1 mm thickness. In the next stage of the test, the top lid was changed to one of the following materials and setup: (1) same mesh sheet as the cylinder, (2) without a lid, (3) copper plate lid and (4) titanium plate lid. The distance between the sample (flat) surface and the top lid was 12 mm. The nearest distance from the mesh cylinder to the sample edge was about 28 mm. A conventional DC plasma nitriding also was done to compare the differences.

The "edge effect" phenomenon was reported in visual inspection of DC plasma nitrided samples and the width of the discolored ring was about 1 to 2 mm on the disc of 25 mm diameter. In contrast, the active screen plasma nitrided sample showed a uniform gray surface through the entire sample surface.

The compound layer in the active screen plasma nitrided sample was reported to be slightly thicker than the layer formed in DC nitrided sample, however, they were equally dense. The morphology of the diffusion zone was lath martensite structure surrounded by the grain boundary and dispersed fine nitrides. All in all, the microstructure variation between the active screen and DC plasma nitrided samples was very little.

The nitrogen depth profiles of DC and active screen plasma nitrided samples are displayed in Fig. 4.13. As shown, the profiles for both methods are similar with a region near the surface with high nitrogen content following with slight reducing underneath. However, the significant difference is that the nitrogen content in the active screen nitrided surface was higher than that in the DC nitrided surface up to a depth of about 6 μm. The high concentration of nitrogen was owing to the existence of γ'-Fe_4N and/or ε-$Fe_{2-3}N$ nitride phases in the compound layer. In fact, the thickness and the phase composition of the compound layer can be estimated from the nitrogen profiles. The slightly deeper high nitrogen profile for active screen plasma nitrided sample in Fig. 4.13 suggested the production of a slightly thicker compound layer in the case of active screen plasma nitrided samples.

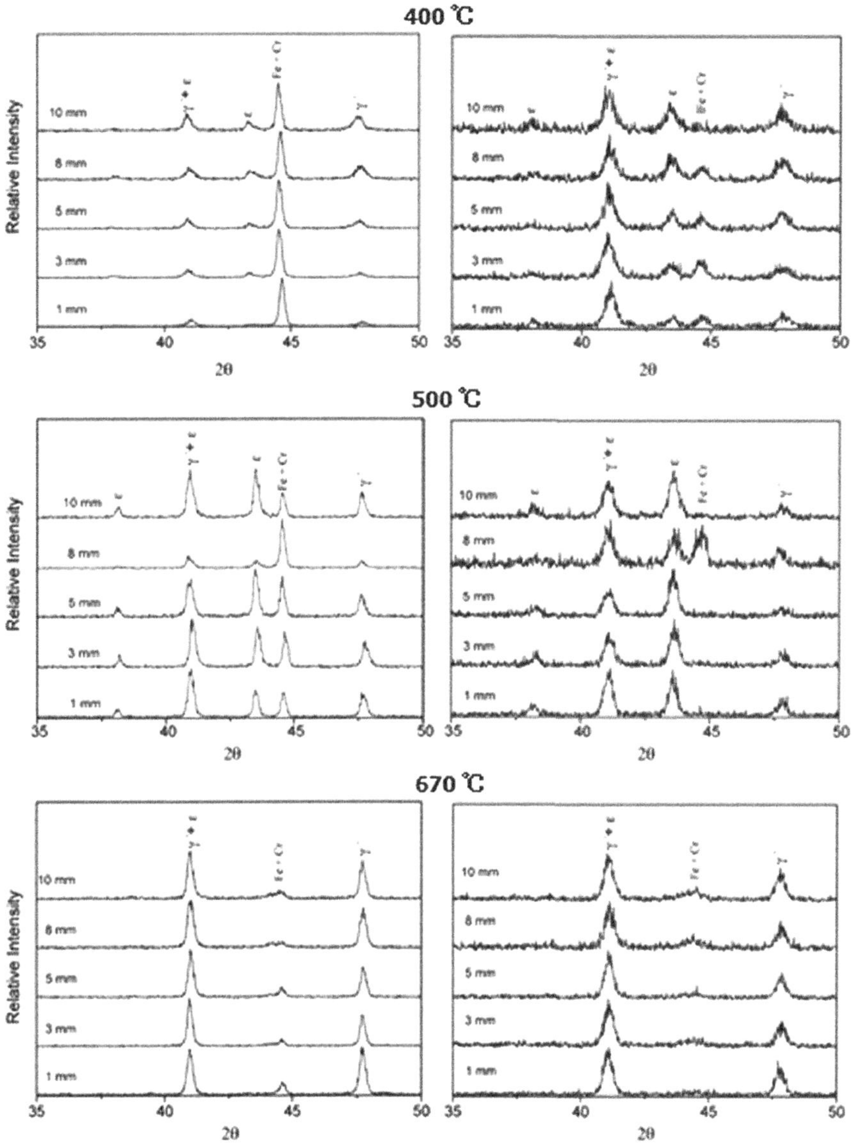

Fig. 4.12 XRD patterns of the samples with 1, 3, 5, 8 and 10 mm of height nitrided in 360 Pa, 80 % of N_2 and temperatures of 400, 500, and 570 °C [22]

The presence of γ'-Fe_4N and ε-$Fe_{2-3}N$ was reported for both active screen and DC plasma treated samples surfaces.

It was observed that active screen plasma nitriding using N_2 + Ar atmosphere produced a nitrogen profile similar to that produced with using $N_2 + H_2$

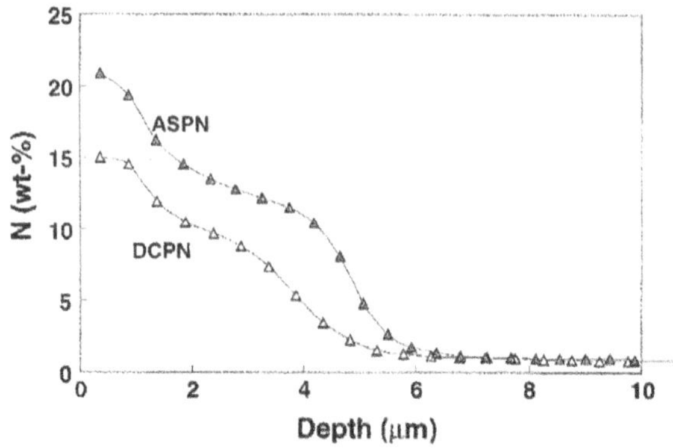

Fig. 4.13 Nitrogen distribution in near surface of active screen and DC plasma nitrided 722M24 steel [23]. (*A study of active screen plasma nitriding, Li C X, Bell T, Dong H, Surface Engineering, Copyright © The Institute of Materials and Mining, reprinted by permission of Taylor & Francis Ltd,* www.tandfonline.com *on behalf of The Institute of Materials, Minerals and Mining*)

atmosphere. Microstructural observations showed a little change in thickness of the compound layer by using N_2 + Ar. However, using 100 % N_2, led to the diffusion of nitrogen only to a very shallow depth below the surface and consequently the presence of much lower nitrogen. No compound layer formation was reported on the sample surface.

It was shown that using mesh lid can produced a similar nitriding effect as that with a steel plate. The dominant constituent of the compound layer on the surface was ε phase but with relatively lower amount.

With removing upper lid, a nonuniform surface was produced by active screen plasma nitriding. Toward the edge and near the mesh cylinder, the sample surface was dark gray, which had a appearance similar to samples nitrided with a steel mesh or a steel plate top lid.

Using a titanium plate as the top lid in active screen method led to the color changing of the nitrided surface to golden mixed with purple probably due to the formation of titanium nitride.

With using copper plate as the top lid, the color of active screen nitrided sample surface turned in reddish brown which was due to the covering of the surface with copper layer.

As reported by Ahangarani et al. [19], the morphologies of DC and active screen plasma nitrided samples were different in size and distribution as seen in Fig. 4.14.

The properties of these particles such as existence, morphology and distribution were dependent on the active screen nitriding conditions. When using argon in nitriding atmosphere, a particle-like morphology similar to Fig. 4.14b was produced although the average size of the particles was a little bigger. With using pure nitrogen, the original grinding marks were remained with seldom particles formed. Without using top lid, a nonuniform layer was deposited on the surface.

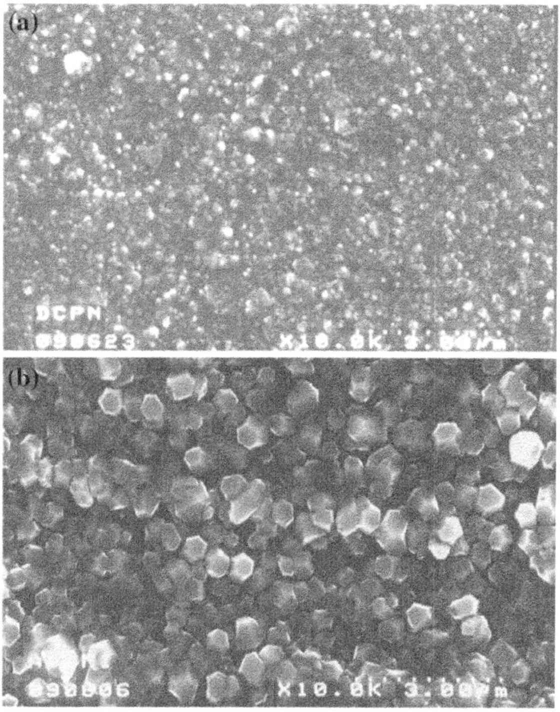

Fig. 4.14 SEM micrographs of **a** DC and **b** active screen nitrided sample surface (nitriding conditions: 500 °C in 25 % N_2 + 75 % H_2 for 5 h; steel plate as top lid in active screen) [23]. (*A study of active screen plasma nitriding, Li C X, Bell T, Dong H, Surface Engineering, Copyright © The Institute of Materials and Mining, reprinted by permission of Taylor & Francis Ltd,* www.tandfonline.com *on behalf of The Institute of Materials, Minerals and Mining*)

Zhao et al. [24] studied the effects of different arrangements (Fig. 4.15) on the active screen plasma nitriding response of the 722M24 low alloy steel.

In Fig. 4.15a, the nitriding treatment was done in a way that the samples and furnace wall were connected to cathodic and anodic potential, respectively similar

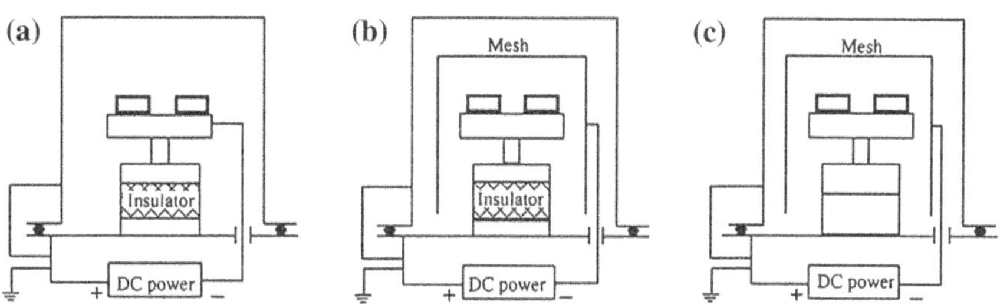

Fig. 4.15 Schematic diagram of plasma nitriding setup showing samples (indicated as 1), **a** in cathodic potential (DC plasma nitriding), **b** in floating potential (AS_f) and **c** in anodic (zero) potential (AS_e) [24]

to DC plasma nitriding. In Fig. 4.15b, a mesh steel cage with diameter and height of 120 and 130 mm was located in the vacuum chamber and was in cathodic potential. However, the samples were in a floating potential and isolated from the cathodic mesh screen and the anodic furnace wall with ceramic and mica pieces similar to the active screen plasma nitriding. In Fig. 4.15c, the samples as well as the furnace wall and the base plate, were earthed and played the role of anode in the DC plasma nitriding system. The mesh steel cage was connected to cathodic potential like that in Fig. 4.15b.

Plasma nitriding was done at 520 °C, for 12 h, under pressure of 5 mbar and atmosphere of 25 % N_2 + 75 % H_2.

They did not see any clear differences in microstructure and thickness of diffusion layer between DC and active screen plasma nitrided samples.

The difference between morphologies of plasma nitrided samples with two methods was also reported as indicated by other authors mentioned previously. However, no clear difference was reported between the surface morphology of the sample nitrided in floating potential and the sample nitrided in anodic potential.

Figure 4.16 shows that the γ'-Fe_4N and ε-$Fe_{2-3}N$ nitride phases are the main constituents of surface layers of 722M24 steel samples produced by both DC and active screen plasma nitriding methods. However, the relative intensities of these two phases are different for DC and active screen plasma nitrided samples. The amount of γ' in the compound layer of DC plasma nitrided sample was relatively higher, while the compound layer on the surface of sample nitrided in floating potential had higher amount of ε.

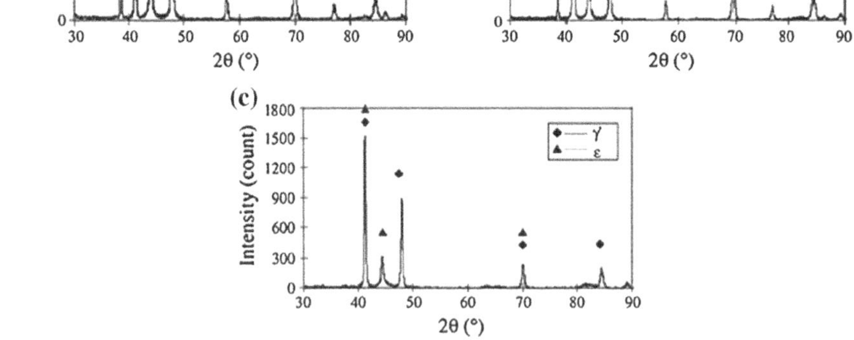

Fig. 4.16 XRD patterns of the DC and active screen plasma nitrided 722M24 steel. **a** DC plasma nitriding, **b** active screen plasma nitriding (floating potential), **c** active screen plasma nitriding (grounded) [24]

It was reported that the order of the nitrided layer thickness from thick to thin was DC plasma nitriding, active screen plasma nitriding at floating potential and active screen plasma nitriding at anodic (zero) potential, respectively. Nevertheless, the difference between the thickness of the layers produced by DC plasma nitriding and the active screen method was small, e.g. less than 10 %.

This study had an important conclusion. It was confirmed that it is not necessary to have sputtering on the surface of the sample to be nitrided. Because nitriding can be achieved and successfully done by placing the sample in either cathodic, floating or anodic (grounded) potential.

Based on these results, the authors also found some phenomena different from the "sputtering and re-condensation" model, as suggested by Kölbel for DC plasma nitriding. In the original Kölbel's model, the iron atoms are sputtered from the surface of cathodic component into the plasma atmosphere. These free iron atoms then react with active nitrogen in the plasma to form FeN phase followed by back-deposition of FeN on the surface of cathodic component. Since FeN is metastable, it changes to $Fe_{2-3}N$ and then Fe_4N nitrides on the steel surface which is associated with releasing of nitrogen for subsequent diffusion.

However, the usefulness of this model was taken under question by the authors. It was claimed that the model may be valid at the beginning of a nitriding process in DC plasma nitriding or for a fresh steel active screen in active screen plasma nitriding. Because, soon after a nitriding process starts, the cathodic surface is covered by a compound layer containing both Fe_4N and $Fe_{2-3}N$ under normal nitriding conditions. Therefore it should be noted that the sputtering target, and thus, the sputtering product are more likely to be iron nitride.

It was found that the sputtered particles are Fe_xN ($x > 2$). Besides, a particular surface morphology was observed for both samples treated in floating potential and anodic potential. These nitride particles which had a hexagonal shape and submicron size, also had a large ratio of superficial area to volume and consequently were very active. Therefore, the mechanism presented as follows: when the sputtered particles are passing through the plasma atmosphere, they would physically and chemically adsorb the active nitrogen species existing in the plasma atmosphere. After deposition of the resultant nitrogen-rich particles on the surface of component, desorption and releasing of physically adsorbed nitrogen atoms would occur as a result of the contact between the particles and the substrate surface and the high substrate temperature. The released nitrogen atoms diffuse into the steel matrix at the nitriding temperature. Part of the particles which adsorbed nitrogen atoms would chemically combine to form nitride, e.g., $Fe_{2-3}N$. This mechanism was named as "sputtering–adsorption–desorption" model as shown in Fig. 4.17. Apparently, the focus of this model is on the importance of the sputtered particles in nitrogen mass transfer, and considering the particles as the main nitrogen carriers in active screen plasma nitriding.

Fig. 4.17 Model of the proposed mechanism for active screen plasma nitriding [24]

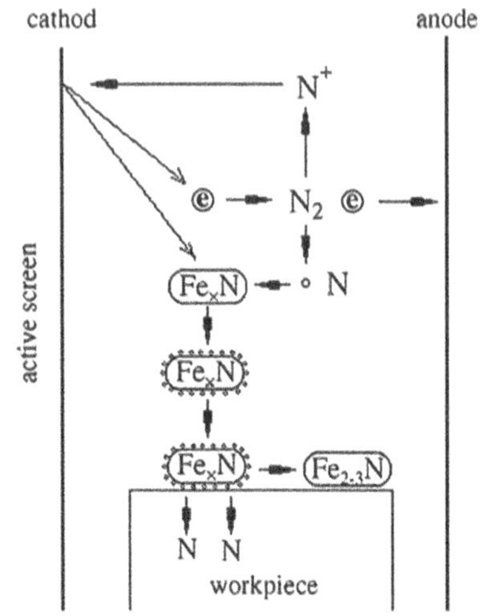

Li et al. [25] investigated active screen plasma nitriding of AISI 5140. The active screen used, consisted of two low carbon steel cylinders with different diameters. The diameter of inner cylinder was 440 mm and that of the outer one was 468 mm. The thickness of the cylinder was 6 mm. The holes of 10 mm diameter were well distributed in the inner cylinder and acted as good entries for the transfer of chemical elements. The distance between two cylinders was 8 mm. The discharge current and voltage between the active screen (composed of two cylinders) and the furnace wall (as anode) was 4–6 A and 500–670 V, respectively. The nitriding atmosphere was NH_3, and the pressure was 150 Pa. The nitriding temperature was between 450 and 550 °C and the process duration was 2, 4, and 6 h. Some of samples were connected to a floating potential while the others connected to a relative lower negative bias voltage of 50 V.

Figure 4.18 shows the typical optical micrographs of the cross section of biased and unbiased samples nitrided at different temperatures during 4 h. As seen in this figure, increasing the process temperature resulted in the increment of compound layer thickness for both of biased or unbiased samples. However, it can be observed that the compound layer obtained for biased samples was much thicker than that of unbiased sample. For example at temperature of 550 °C, a continuous compound layer of ~5.0 thick μm was formed on the unbiased sample (Fig. 4.18e), while in the case of biased sample, the layer was relatively thicker (~7.0 μm) (Fig. 4.18f).

It was reported that at lower treatment time and temperature, small amounts of ε-$Fe_{2-3}N$ and γ'-Fe_4N were formed on the surface. However, increase in the treatment temperature and time resulted in an increase in the amount of the ε-$Fe_{2-3}N$ and γ'-Fe_4N phases for both biased and unbiased samples. Thus, it was concluded that the formation of ε-$Fe_{2-3}N$ nitride phase was facilitated by plasma nitriding in the biased samples.

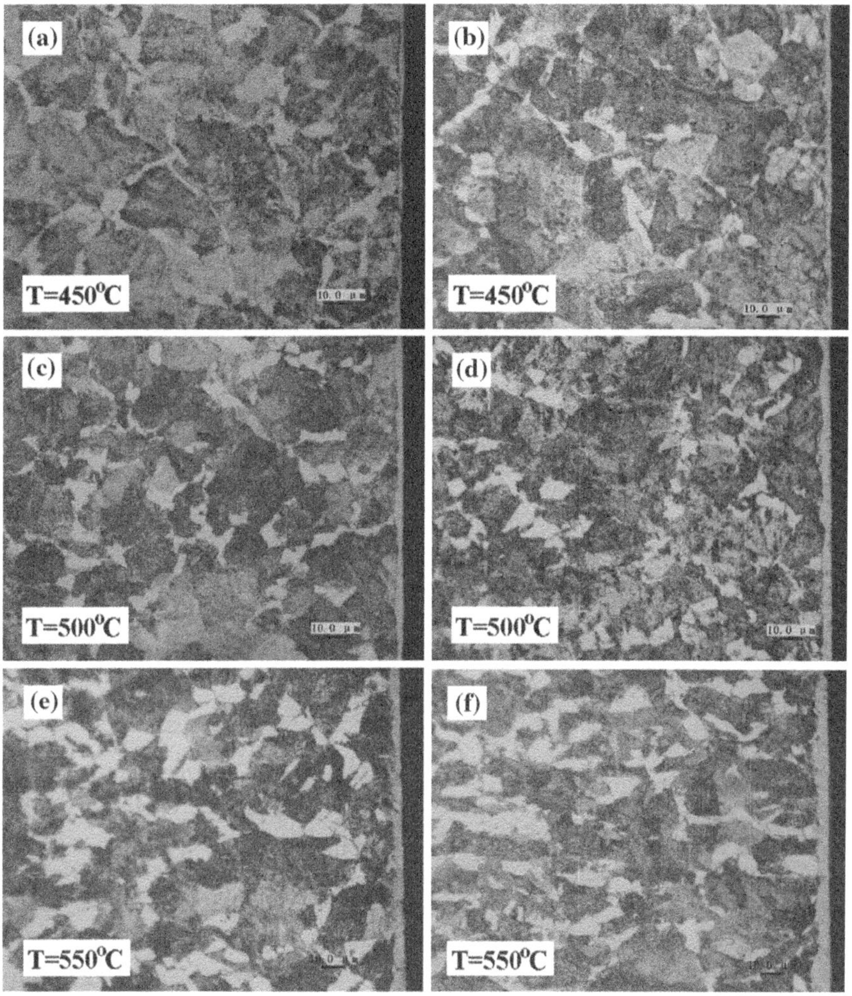

Fig. 4.18 Optical cross-sectional images of samples nitrided with bias voltage of **a**, **c**, and **e** 0 V and **b**, **d**, and **f** −50 V [25]

4.2 Mechanical Properties

4.2.1 Austenitic Stainless Steels

Gallo et al. [10] (see Sect. 4.1.1) observed very clear difference between the microhardness values corresponding to the area of the sample under the hole and the area of the sample covered by the mesh (Fig. 4.19). This figure clearly shows a considerable hardening effect for the area covered by the mesh in comparison with the area under the hole. However, from visual point of view, there was a little

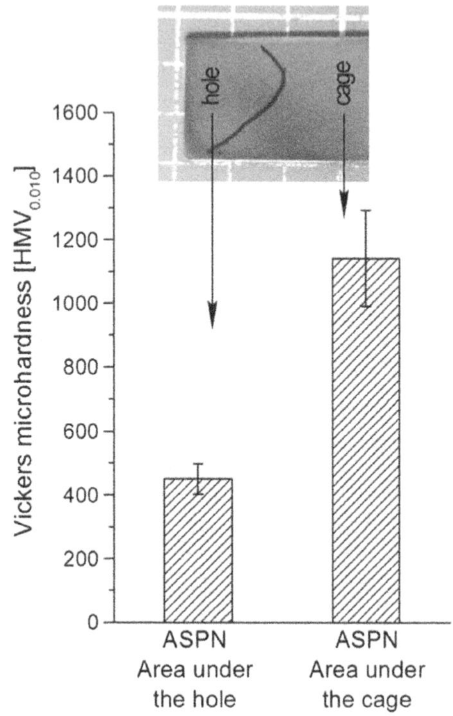

Fig. 4.19 Vickers microhardness values measured on different areas of the same AISI 304 sample nitrided using active screen method [10]

difference between their appearances. The area covered by the mesh became dull gray while the under-hole area became shiny due to the deposition layer.

Alves Jr. et al. [12] (see Sect. 4.1.1) reported the differences in hardness values between erosion rings (where the "edge effect" occurs) and central area. Figure 4.20 presents the results of microhardness measurements, with three indentation points for each sample, in both active screen and DC plasma nitrided samples. In the DC plasma nitriding technique, a reduction in hardness of the erosion ring region, compared with the central area, was observed. Also it is obvious that, because of the thermal gradient, the microhardness value in edge areas is much higher than the central area. Therefore, this difference caused the coating to break along the edge line. In contrast, for active screen plasma nitrided samples, the hardness values remained uniform in the whole sample and presented the same order of magnitude that the central region of DC plasma nitrided samples.

Li et al. [13] (see Sect. 4.1.1) reported a significant improvement in surface hardness as well as load bearing capacity after active screen plasma nitriding treatment.

For samples nitrided at 420 °C using DC and active screen methods, indentation test was done. It was reported that, by reaching the indentation load to a critical value, cracks started to appear around the indentation mark and propagate with further increasing the load. The above-mentioned critical value is as 0.2 kg for DC

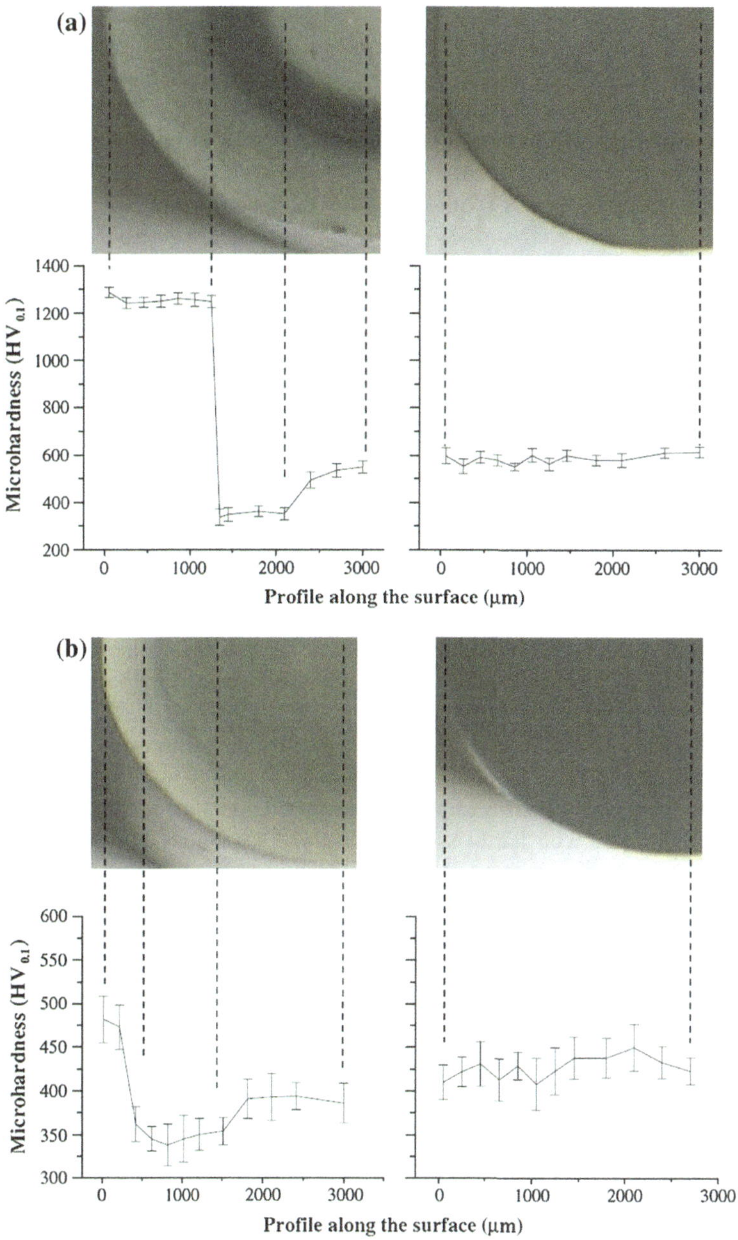

Fig. 4.20 *The upper part* top view of the samples. *The lower part* radial microhardness values in the *top* surface of the samples **a** DC plasma nitrided and **b** active screen plasma nitrided [12]

and 0.5 kg for active screen method. Therefore, it can be concluded that the lower temperature active screen plasma nitriding resulted in better ductility and higher load bearing capacity.

The different ductility values obtained by active screen and DC plasma nitriding was ascribed to the different outer surface morphologies resulted by these two methods. It was revealed that the surface of the sample treated by DC plasma nitriding, consisted of "cones" with irregular sizes and distributions resulted by the sputtering effect. However, surface of the sample treated by active screen plasma nitriding, consisted of multiple particles with clear boundaries which was believed to be the result of the deposition effect.

In the case of tribological properties, using a pin-on-disc tribometer and sliding distance of 500 m, a slight variance was observed in the wear rates of the nitrided samples with the nitriding temperature, as well as with the composition, the thickness and the hardness of the surface layer. However, plasma nitriding significantly reduced the wear rate of all treated samples in comparison with the untreated ones. The wear rate of the nitrided samples was two orders of magnitude less than the untreated samples under the present testing conditions.

However, no considerable difference was found between wear resistance of the samples treated with active screen plasma nitrided samples and DC plasma nitrided samples.

The nitrided layer thickness and microhardness values obtained for AISI 316 austenitic stainless steel by de Sousa et al. [14] (see Sect. 4.1.1) are presented in Table 4.3. The thickness of the nitrided layer increased with nitrogen content up to 80 % but decreased when 100 % N_2 was used as nitriding atmosphere. Furthermore, higher microhardness was obtained for the sample nitrided in 80 % nitrogen. It was claimed that contrary to other researchers, using 100 N_2 led to the formation of a layer with hardness significantly greater than that of the substrate when using the mentioned treatment conditions.

Li et al. [15] (see Sect. 4.1.1) observed that 800 m sliding produced a wear track with a width and depth of about 2 mm and 100 μm on the unnitrided AISI 316 stainless steel surface as seen in Fig. 4.21a. However, the active screen plasma nitrided samples exhibited significantly less wear rate. The wear tracks on the plasma nitrided surfaces were shallow and superficial as seen in Fig. 4.21b. The wear track on the surface of nitrided sample was still much shallower even after a very long distance sliding (e.g. 2000 m). Beside the improved wear resistance,

Table 4.3 Average thickness and microhardness values of the nitrided layers formed on AISI 316 samples (Nitriding treatments were performed at 250 Pa and 450 °C for 5 h with different nitriding atmospheres) [14]

Nitriding mixtures (%N_2)	Layer thickness (μm)	Microhardness (HV)
20	3.6	530
50	5.6	460
80	31.6	1100
100	21.0	950

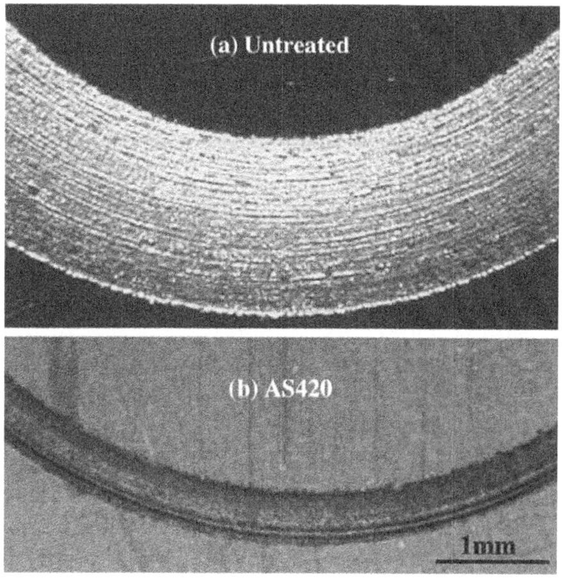

Fig. 4.21 Wear tracks on: **a** unnitrided; **b** active screen plasma nitrided at 420 °C AISI 316 austenitic stainless steel sample surface (sliding for 500 m, contact load 10 N) [15]

active screen plasma nitriding also reduced the friction coefficient of AISI 316 sample against WC/Co from an average value of 0.4–0.2.

Figure 4.22 shows the variation of wear volume loss (mm^3) with sliding distance for the unnitrided and active screen plasma nitrided AISI 316 sample. As can be seen in this plot, the volume loss of the active screen plasma nitrided sample was more than two orders of magnitude less than that of the unnitrided sample. The wear resistance improvement was similar for the active screen plasma nitrided samples at 420 and 500 °C. This confirmed the significant improvement of sliding wear resistance for active screen plasma nitrided AISI 316 sample.

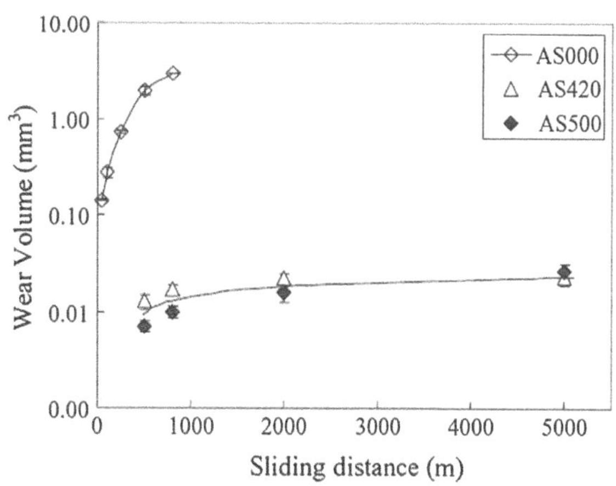

Fig. 4.22 Variations of wear volume loss with sliding distance for unnitrided and active screen plasma nitrided 316 austenitic stainless steel samples at different temperatures [15]

The material loss increased with sliding distance for all the samples, however, the wear rate (mm^3/m) showed different behavior.

They reported the increment of wear rate of the unnitrided sample with sliding distance until a maximum of 3.94×10^{-3} mm^3/m at 500 m and its decrement to 3.7×10^{-3} mm^3/m at 800 m. The wear rate of nitrided samples (which was much lower than unnitrided samples) was observed to reduce with increasing sliding distance from the beginning of the test. For example, the wear rate of the sample nitrided at 420 °C was reduced from 2.54×10^{-4} mm^3/m at 500 m to 0.46×10^{-5} mm^3/m at 5000 m, the latter being about one fifth of the former. At short sliding distance, the wear rate of sample nitrided at 500 °C was lower than that of sample nitrided at 420 °C, but the difference decreased with increasing sliding distance. After 5000 m sliding, the wear rate of the samples nitrided at higher temperature was almost the same as that for the samples nitrided at lower temperature.

It was concluded that the wear of the untreated AISI 316 austenitic stainless steel occurred by a combined mechanism of adhesion, abrasion, and plastic deformation.

The surface of unnitrided AISI 316 austenitic stainless steel experienced a severe deformation during wear test due to its relatively low hardness. It was suggested that the combination of strong adhesion, abrasion, and severe plastic deformation corresponded for the very high and continuously increased wear rate of the unnitrided AISI 316 stainless steel as shown in Fig. 4.22. Likely, the initial wear rate increased with sliding distance due to the gradual increment of contact area and enhanced occurrence of adhesion. After sliding for a long distance, a reduction was observed in wear rate. This behavior was ascribed to the reduction of contact stress as a result of increased contact area and the high amount of wear debris existing in the deep wear track.

Active screen plasma nitriding caused the surface hardness of AISI 316 sample to be enhanced from ~200 HV to more than 1000 HV. This hardening effect changed the wear mechanism of the steel during sliding. It was observed that, the wear tracks on surface of all the nitrided samples were shallow, superficial and very smooth. Furthermore, features of adhesion and plastic deformation, (e.g., the deep abrasive grooves, extensive smearing of materials and severe shearing deformation) were not observed. Therefore, the worn surface of plasma nitrided and unnitrided samples was completely different.

It was also reported no clear difference between the wear resistance of DC and active screen plasma nitrided samples.

The addition of CH_4 to nitriding atmosphere resulted in a further improvement in mechanical properties of active screen plasma nitrided AISI 316 austenitic stainless steel as reported by de Sousa et al. [17] (see Sect. 4.1.1). The microhardness depth profiles presented in Fig. 4.23 show an abrupt reduction with depth (most clearly shown in Fig. 4.23b), when compared with the active screen plasma nitrided samples without CH_4 addition as shown in Fig. 4.24. This feature of active screen plasma nitrided samples, led to less possibility of the layer detachment. Additionally, the properties remained constant along the whole surface of sample and the so-called "edge effect" was almost entirely eliminated.

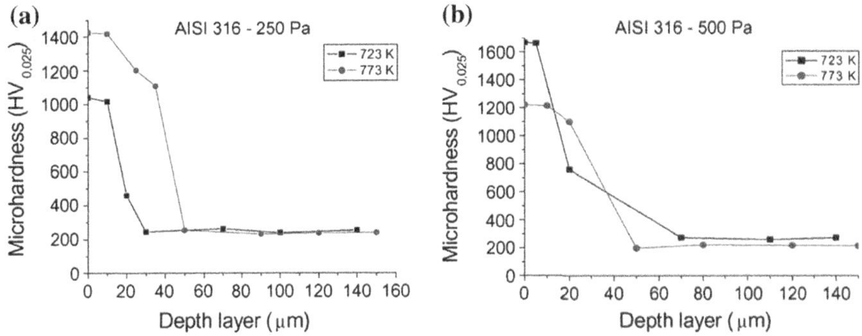

Fig. 4.23 Variation of microhardness with depth for samples nitrided at **a** 250 Pa and **b** 500 Pa [17]

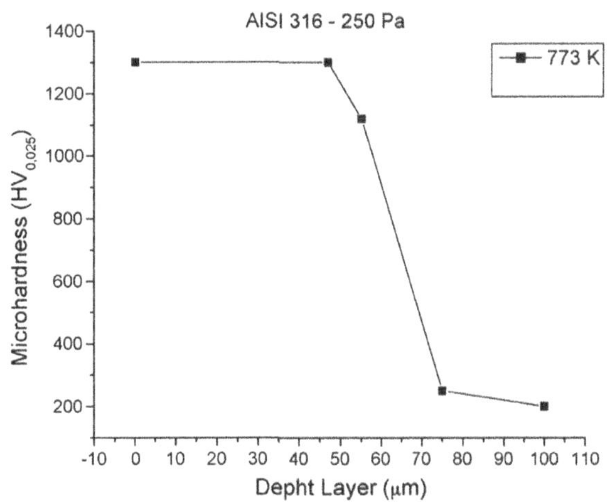

Fig. 4.24 Microhardness variations with depth for samples nitrided by active screen method at temperature of 500 °C, without CH_4 addition [17]

Nishimoto et al. [18] investigated the effect of the distance between screen and sample on hardness of active screen plasma nitrided low carbon S15C steel and SUS 304 austenitic stainless steel.

The surface hardness increment was reported with decreasing the distance between sample and screen due to the thickening of the modified layer. For low carbon steel, a lower hardness for diffusion layer was also reported with increasing distance. This happened because of lower diffusion of nitrogen toward the core.

4.2.2 Structural and Constructional Steels

Ahangarani et al. [19] (see Sect. 4.1.2) reported the microhardness improvement after active screen plasma nitriding of 30CrNiMo8 low alloy steel. It was reported that all samples showed high surface microhardness values following by a drop at the interface of case and core to that of substrate. Higher surface hardness values were obtained for active screen plasma nitrided samples at 75 % N_2 + 25 % H_2 gas mixture and 580 °C at 5 h of treatment time. It was observed that microhardness values depended on the treatment duration. This was ascribed to the dispersion of nitride precipitates in the matrix. The dislocations movement was obstructed by presence of precipitates with a specific size leading to the most strengthening and hardening effect. With prolonging the treatment, the precipitates particles got longer in size and more prone to coarsening, resulting in a lower precipitate density and consequently lower hardness.

Also the microhardness values obtained by using screen top lids were slightly higher than that obtained by iron plate top lid. This was likely due to the effective role of the holes of top lids in facilitating of ions transition from plasma atmosphere (outside of the screen) onto the surface of sample (inside of the screen). Therefore, the best surface hardening effect was obtained with increasing treatment temperature and using screen top lid.

de Sousa et al. [22] observed that the microhardness measured along the nitrided surface of samples with different heights treated at the same treatment temperature, shows a complete uniformity as it is presented in Fig. 4.25 for the case of 500 °C.

Li et al. [23] (see Sect. 4.1.2) investigated the different active screen plasma nitriding parameters on the microhardness obtained. The average surface hardness was 1025 $HV_{0.1}$ for active screen plasma nitriding and 1038 $HV_{0.1}$ for DC plasma nitrided samples. The hardness profiles and the case depth produced by two nitriding processes practically had little difference.

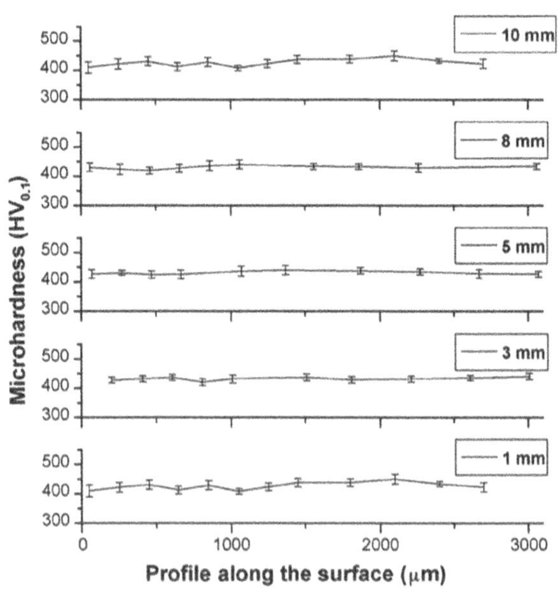

Fig. 4.25 Microhardness in the top surface of the samples nitrided at 500 °C [22]

Hardness measured on the surface of the sample treated using Ag + N_2 was slightly lower than active screen nitriding with N_2 + H_2 (Fig. 4.26), but the hardness measured in the diffusion zone was almost the same. It can be seen that active screen plasma nitriding using N_2 + Ar can achieve a similar hardening effect as using N_2 + H_2. However, when pure nitrogen was used, nitrogen diffused only to a very shallow depth below the surface, and the nitrogen content was also much lower. No compound layer observed when using pure nitrogen. The surface hardness of the active screen plasma nitrided sample was nearly the same as the untreated steel (Fig. 4.26), because no hardening effect was found below the surface. Therefore, it was concluded that active screen plasma nitriding using pure nitrogen did not produce any noticeable hardening.

It was found that using mesh lid can produce a similar nitriding effect as that with a steel plate, including a similar average surface hardness on the surface and a similar hardness profile in the nitrided case.

A very small hardening effect was observed by using Ti top lid. In the case of Cu top lid, there was not any hardening at all. This could be owing to the deposition of titanium and copper and not allowing nitrogen to reach the sample surface.

Zhao et al. [24] (see Sect. 4.2.2) concluded that, different nitriding conditions depicted in Fig. 4.15, had a little effect on the surface hardness of 722M24 steel. It was observed that the surface hardness of the untreated sample was around 320 $HV_{0.1}$. After plasma nitriding, the hardness increased to more than 1000 $HV_{0.1}$.

Li et al. [25] (see Sect. 4.2.2) observed that, increasing treatment temperature resulted in hardness increment, however, there was no difference between biased and unbiased samples in this case. They also reported higher surface hardness for biased samples compared to unbiased ones. The measured hardness values for biased samples were 620 HV, 835 HV, and 940 HV for the 450, 500, and 550 °C treatment temperatures respectively.

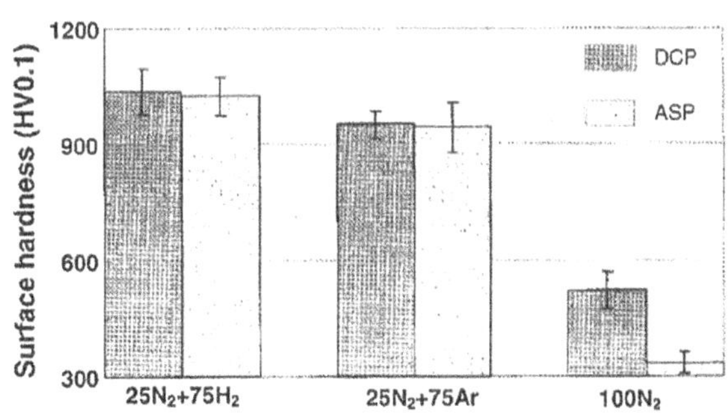

Fig. 4.26 Effect of gas composition on surface hardness of active screen and DC nitrided samples [23]. (*A study of active screen plasma nitriding, Li C X, Bell T, Dong H, Surface Engineering, Copyright © The Institute of Materials and Mining, reprinted by permission of Taylor & Francis Ltd,* www.tandfonline.com *on behalf of The Institute of Materials, Minerals and Mining*)

4.3 Corrosion Properties

4.3.1 Austenitic Stainless Steels

Li et al. [16] (see Sect. 4.1.1) investigated the effects of active screen plasma nitriding on corrosion properties of AISI 316 austenitic stainless steel. First of all, the immersion test was done to evaluate corrosion properties. With etching in 50 % HCl + 25 % HNO_3 + 25 % H_2O solution, the sample nitrided at 420 °C showed uniform corrosion with bright appearance indicating low corrosion. However, the corrosion of untreated sample was observed to be ununiform due to the existence of grain boundaries and twin boundaries. For the sample nitrided at 500 °C, a dark appearance was observed after immersion test that was due to its poor corrosion resistance.

These results were confirmed by another immersion in 10 % HCl water solution. As seen in Fig. 4.27, the weight loss and consequently corrosion rate was higher for the sample nitrided at 500 °C with a factor of 5–10. The surface of sample after test, was dark and full of dark corrosion products.

Also, it can be inferred from Fig. 4.27 that, low temperature treatment did not improve corrosion resistance but kept it similar to untreated condition. It was reported that, in the case of plasma nitriding at 420 °C, general corrosion occurred instead of pitting corrosion occurred for untreated sample.

The corrosion response in electrochemical tests also was evaluated. For untreated sample, a sudden increase in the current density was observed at potential of almost 0 V due to the break down of the passive film formed on the surface. Consequently, localized corrosion occurred. For untreated sample, the main corrosion types were reported as pitting and crevice corrosion.

With nitriding at 420 °C, the corrosion potential increased from −208.8 to −83.8 mV. Also, the current density decreased from 1.7 to 0.1 mA/cm^2. However,

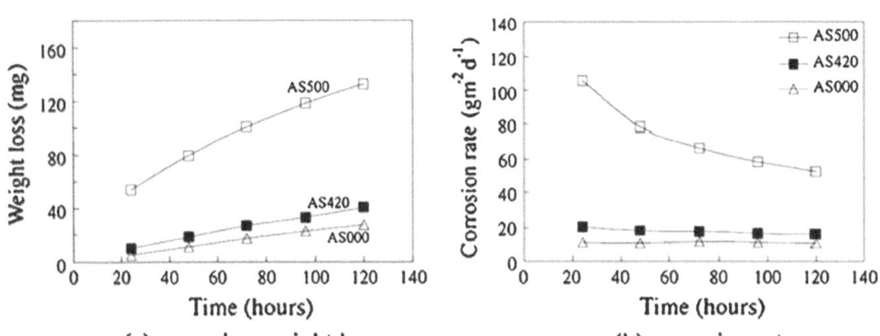

Fig. 4.27 Corrosion weight loss and corrosion rate of AISI 316 stainless steel samples nitrided at various temperatures in 10 % HCl water solution [16]

due to the slow increment of dissolution current density with potential, it was concluded that the general corrosion occurred.

All in all, from the above results, it can be seen that active screen plasma nitriding at 420 °C improved corrosion resistance compared to untreated sample.

As seen before in Fig. 4.27, active screen plasma nitriding at 500 °C resulted in deteriorating the corrosion resistance of AISI 316. Figure 4.28 shows the morphologies of the corroded surface of the sample nitrided at 500 °C. It is obvious from this figure that the corrosion in this sample was localized but uniform. Also two distinct areas are obvious in Fig. 4.28a: bright and dark. Close examinations revealed that the morphology and chemical composition of bright area are the same as for as-nitrided case and the corrosion attacks did not occur or occurred in negligible amount in these areas.

However, it can be indicated from Fig. 4.28 that the dark area can be a result of the removal of a layer on the surface because of the dissolution or spallation (Fig. 4.28b). Corrosion pits and corroded grain boundaries are also evident in Fig. 4.28c.

It can be concluded that, similar to previous methods, with formation of CrN in nitriding at higher temperature, chromium in the matrix was depleted around the nitrided precipitates and consequently the formation of a dense and continuous

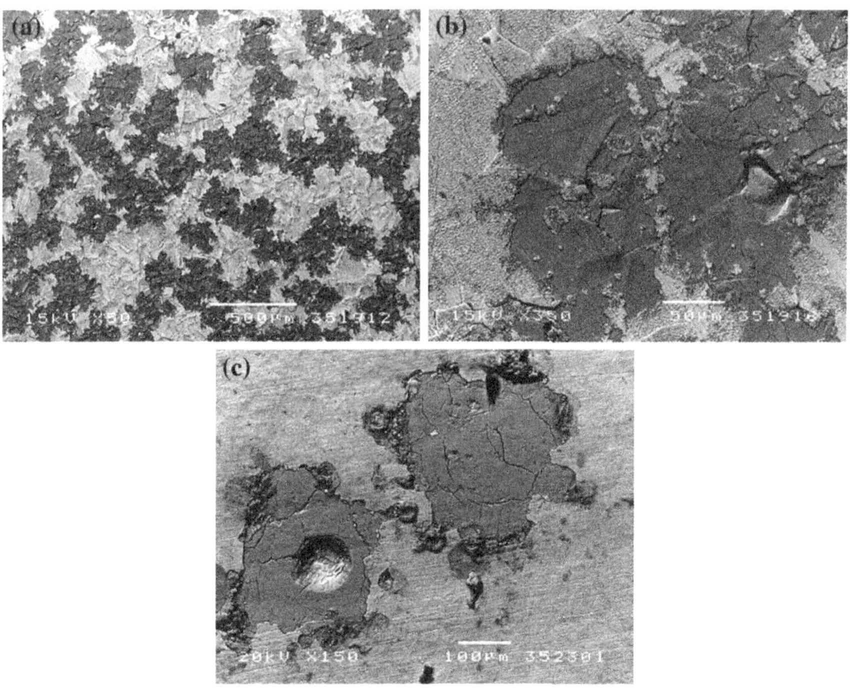

Fig. 4.28 Corroded surface morphologies of the sample plasma nitrided at 500 °C using active screen technique [16]

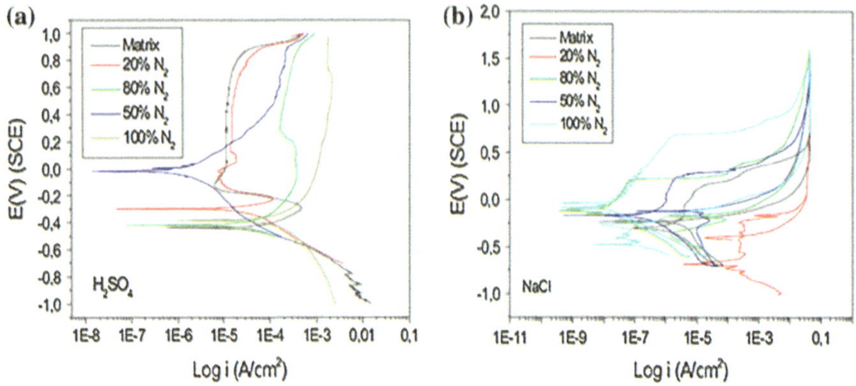

Fig. 4.29 Anodic polarization curves for untreated and plasma nitrided AISI 316L samples, **a** in 0.1M H_2SO_4 solution and **b** in 3.5 wt% NaCl solution [14]

oxide layer on the surface was prevented. This led to the deterioration of the corrosion resistance.

de Sousa et al. [14] (see Sect. 4.1.1) found that in contrast to previous investigations, compound layer formed in 100 % N_2 atmosphere had a poor corrosion resistance in H_2SO_4 solution and a moderate corrosion resistance in NaCl solutions (as obvious from polarization curves in Fig. 4.29).

In case of 0.1M H_2SO_4 solution, the samples treated in 50 and 100 % N_2 atmosphere showed low corrosion potential as observed for untreated sample. The passive current density of these samples was two orders of magnitude larger than that of the matrix and the sample nitrided at 20 % N_2 showing the difficulty of passive layer to form in pure 100 % N_2 atmosphere. However in case of 3.5 % NaCl solution, a considerable change in the corrosion resistance was observed for all conditions. The lowest corrosion current density among all nitrided samples belonged to the sample nitrided in 100 % N_2 atmosphere.

4.3.2 Structural and Constructional Steels

Ahangarani et al. [20] (see Sect. 4.1.2) observed the improvement of corrosion resistance after both active screen and DC plasma nitriding for 30CrNiMo8 low alloy steel. The polarization curves are presented in Fig. 4.30. It can be observed that in DC plasma nitriding, increasing treatment temperature resulted in increasing of corrosion potential (E_{corr}), decreasing of corrosion current density (I_{corr}) and consequently corrosion rate. This means that in this case, nitriding was capable of improving corrosion resistance.

As can be seen in Fig. 4.30 the corrosion potential shifted to higher values at 550 °C and atmosphere of 75 % N_2 + 25 % H_2 after plasma nitriding. However, it

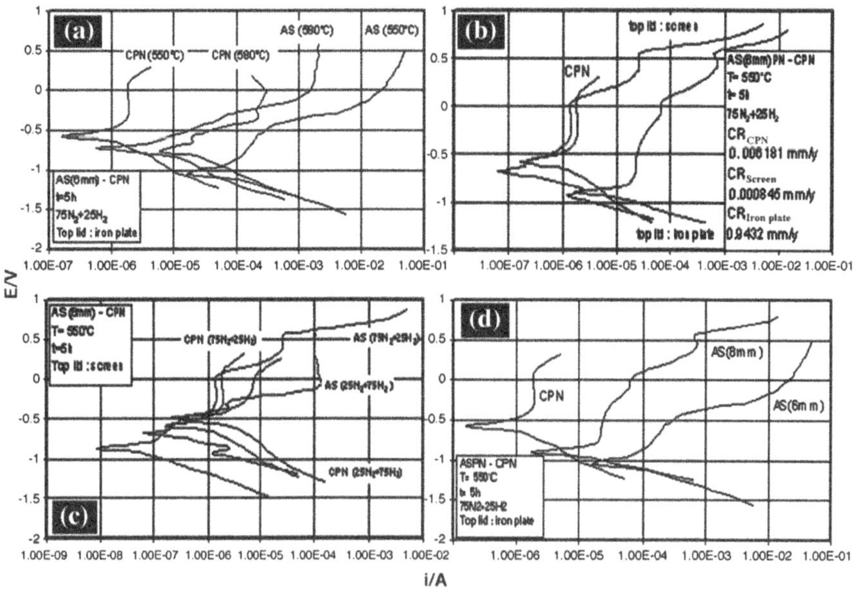

Fig. 4.30 Polarization curves for 30CrNiMo8 low-alloy steel treated with active screen and DC plasma nitriding at 550 and 580 °C for 5 h at various atmospheres, for both top lids and both hole sizes of screen [20]

remained constant at 580 °C. Also, it is obvious that for active screen plasma nitriding, using iron plate lid resulted in increased corrosion potential.

Figure 4.30c shows that for active screen plasma nitriding, the value of E_{corr} had a direct relationship with the amount of N_2 in nitriding atmosphere. However, for DC plasma nitriding this relationship was reverse. Figure 4.30d shows that increasing screen hole size from 6 to 8 mm had no noticeable effect on the potential.

In general, the corrosion potential for both active screen and DC plasma nitriding was higher than for untreated sample.

The improvement in corrosion properties also was reported by Li et al. [25] (see Sect. 4.1.2) for AISI 5140 low alloy steel. For all three treatment temperatures, increasing corrosion potential and decreasing corrosion current density were observed. Although, untreated sample suffered from severe general corrosion, plasma nitriding promoted passivation and consequently corrosion resistance.

The values obtained for the corrosion potential were $-0.20\ V_{(SCE)}$, $-0.19\ V_{(SCE)}$, and $-0.22\ V_{(SCE)}$ for the biased samples nitrided at 450, 500, and 550 °C, respectively. At the same polarization potentials, the current densities of biased samples were much lower than those of unbiased samples.

Therefore, the biased sample nitrided at 550 °C for 4 h showed the highest corrosion resistance. This was probably due to the formation of relatively high amount of ε-$Fe_{2-3}N$ phase during nitriding.

References

1. Soltani Asadi Z, Mahboubi F (2012) Effect of component's geometry on the plasma nitriding behavior of AISI 4340 steel. Mater Des 34:516–521
2. Yazdani A, Soltanieh M, Aghajani H, Rastegari S (2011) A new method for deposition of nano sized titanium nitride on steels. Vac 86:131–139
3. Ribeiro K J B, de Sousa R R M, de Ara′ujo F O, de Brito R A, Barbosa J C P, Alves Jr C (2008) Industrial application of AISI 4340 steels treated in cathodic cage plasma nitriding technique. Mater Sci Eng A 479:142–147
4. Kurelo B C E S, de Souza G B, Rutz da Silva S L, Daudt N de F, Alves Jr C, Torres R D, Serbena F C (2015) Tribo-mechanical features of nitride coatings and diffusion layers produced by cathodic cage technique on martensitic and supermartensitic stainless steels. Surf Coat Technol 275:41–50
5. Lin K, Li X, Su Y, Luo X, Dong H (2014) Active screen plasma nitriding of 316 stainless steel for the application of bipolar plates in proton exchange membrane fuel cells. Int J Hydrog Energy 39:21470–21479
6. (1991) ASM Handbook Vol. 4, Heat Treating. ASM International, United States of America
7. Griin R, Giinther H J, (1991) Plasma nitriding in industry-problems, new solutions and limits. Mater Sci Eng A 140:435–441
8. Gallo S C, Dong H (2012) New insights into the mechanism of low-temperature active-screen plasma nitriding of austenitic stainless steel. Scr Mater 67:89–91
9. Hubbard P, Dowey S J, Doyle E D, McCulloch D G (2006) Influence of bias and in situ cleaning on through cage (TC) or active screen plasma nitrided (ASPN) steels. Surf Eng 22:243–247
10. Gallo S C, Dong H (2010) On the fundamental mechanisms of active screen plasma nitriding. Vac 84:321–325
11. de Sousa R R M, de Arau′jo F O, da Costa J A P, Dumelow T, de Oliveira R S, Alves Jr C (2009) Nitriding in cathodic cage of stainless steel AISI 316: Influence of sample position. Vac 83:1402–1405
12. Alves Jr C, de Araújo F O, Ribeiro K J B, da Costa J A P, Sousa R R M, de Sousa R S (2006) Use of cathodic cage in plasma nitriding. Surf Coat Technol 201:2450–2454
13. Li C X, Georges J, Li X Y (2002) Active screen plasma nitriding of austenitic stainless steel. Surf Eng 18:453–457
14. de Sousa R R M, de Araújo F O, Gontijo L C, da Costa J A P, Alves Jr C (2012) Cathodic cage plasma nitriding (CCPN) of austenitic stainless steel (AISI 316): Influence of the different ratios of the (N_2/H_2) on the nitrided layers properties. Vac 86:2048–2053
15. Li C X, Bell T (2004) Sliding wear properties of active screen plasma nitrided 316 austenitic stainless steel. Wear 256:1144–1152
16. Li C X, Bell T (2004) Corrosion properties of active screen plasma nitrided 316 austenitic stainless steel. Corros Sci 46:1527–1547
17. de Sousa R R M, de Ara′ujo F O, Barbosa J C P, Ribeiro K J B, da Costa J A P, Alves Jr C (2008) Nitriding using cathodic cage technique of austenitic stainless steel AISI 316 with addition of CH_4. Mater Sci Eng A 487:124–127
18. Nishimoto A, Nagatsuka K, Narita R, Nii H, Akamatsu K (2010) Effect of the distance between screen and sample on active screen plasma nitriding properties. Surf Coat Technol 205:S365–S368
19. Ahangarani Sh, Sabour A R, Mahboubi F (2007) Surface modification of 30CrNiMo8 low-alloy steel by active screen setup and conventional plasma nitriding methods. Appl Surf Sci, 254:1427–1435
20. Ahangarani Sh, Sabour A R, Mahboubi F, Shahrabi T (2009) The influence of active screen plasma nitriding parameters on corrosion behavior of a low-alloy steel. J Alloys Compd 484:222–229

21. Ahangarani Sh, Mahboubi F, Sabour A R (2006) Effects of various nitriding parameters on active screen plasma nitriding behavior of a low-alloy steel. Vac 80:1032–1037
22. de Sousa R R M, de Ara´ujo F O, Ribeiro K J B, Mendes M W D, da Costa J A P, Alves Jr C (2007) Cathodic cage nitriding of samples with different dimensions. Mater Sci Eng A 465:223–227
23. Li C X, Bell T, Dong H (2002) A study of active screen plasma nitriding. Surf Eng 18: 174–181
24. Zhao C, Li C X, Dong H, Bell T (2006) Study on the active screen plasma nitriding and its nitriding mechanism. Surf Coat Technol 201:2320–2325
25. Li Y, Wang L, Zhang D, Shen L (2010) Influence of bias voltage on the formation and properties of iron-based nitrides produced by plasma nitriding. (2010) J Alloys Compd 497:285–289

Chapter 5
Radiofrequency (RF) Plasma Nitriding

The usual plasma nitriding is done using glow discharge with the workpiece as cathode at pressures between 100 and 1000 Pa. This procedure leads to bombardment of the surface by energetic ions, therefore the temperature is high.

However, as noted in Chap. 2, high temperature is not sufficient for plasma nitriding of austenitic stainless steels due to the formation of CrN phase which deteriorates the corrosion resistance. For this purpose, some plasma nitriding methods have been introduced at low pressures such as RF, microwave, or electron-beam plasma nitriding.

As well as the possibility of these methods for low-temperature treatments, other advantages can be noted for these methods such as lower cost, less distortion, and less surface roughening. Also, lowering the process temperature widens the range of alloys and components that can be treated.

In these methods and specially RF, the plasma is generated by a source separate from the workpiece at lower pressures (approximately 0.1 Pa). At this low-pressure RF plasma nitriding, the generated plasma diffuses throughout the treatment chamber and contains a large number of active species. Because of the reduced collisionality at the lower pressure, the nitriding efficiency increases in this method [1–3].

The following advantages have been presented for RF plasma nitriding:

1. Using lower temperatures
2. Minimal distortion of the treated workpiece
3. Lower consumption of feed gas due to the low-pressure needed
4. No need for arc suppression due to the separation of plasma generation from substrate bias
5. Better control on substrate temperature
6. Shortening the nitriding time (minutes instead of hours)
7. More slowly falling the thickness of nitrided layer [4–7].

In this chapter, the effects of RF plasma nitriding on microstructural, mechanical, and corrosion properties of steels will be discussed.

H. Aghajani and S. Behrangi, *Plasma Nitriding of Steels*, Topics in Mining, Metallurgy and Materials Engineering, DOI 10.1007/978-3-319-43068-3_5

5.1 Microstructure

5.1.1 Austenitic Stainless Steels

Baldwin et al. [6] plasma nitrided AISI 316 austenitic stainless steel at temperature of 400 ± 10 °C, frequency of 13.56 MHz, under pressure of 36–900 mPa, in 99.99 % N_2 atmosphere (<12 ppm H_2O and <10 ppm O_2), and for 0.75–9 h.

It was observed that increasing treatment time resulted in thicker nitrided layer. Also, they observed the shifting of austenite peaks to lower angles and rising expanded austenite with increasing pressure.

Ibrahim et al. [8] studied the effects of different nitriding times on plasma nitriding behavior of AISI 304 austenitic stainless steel. They did RF plasma nitriding with frequency of 13.6 MHz, at temperature of 625 °C, under atmosphere of pure N_2, pressure of 0.1 torr and for different durations (3, 4, 5, 6, 7, 10, 15, 25, 30, and 40 min).

For short plasma time (5 min), nitride phases of FeNiN and ε-$Fe_{2-3}N$ were reported to be produced. With an increase in nitriding time (7 min), the FeNiN disappeared and replaced with γ′-Fe_4N. With a further increase in the duration of nitriding, the ε-$Fe_{2-3}N$ phase was completely disappeared. At 40 min, the microstructure mainly constituted austenite and CrN phases. Some traces of γ′-Fe_4N also were detected.

Valencia-Alvarado et al. [9] studied plasma nitriding behavior of AISI 304 stainless steel with RF frequency of 13.56 MHz in atmosphere of 85 % H_2 + 15 % N_2. They also biased substrate from 0 to −300 V for 5 h.

Figure 5.1 shows the sample temperature growth rate versus bias voltages (0 to −300 V), in the first 20 min. It was seen that the substrate temperature reached 200 and 470 °C for bias voltages of 0 and −300 V, respectively.

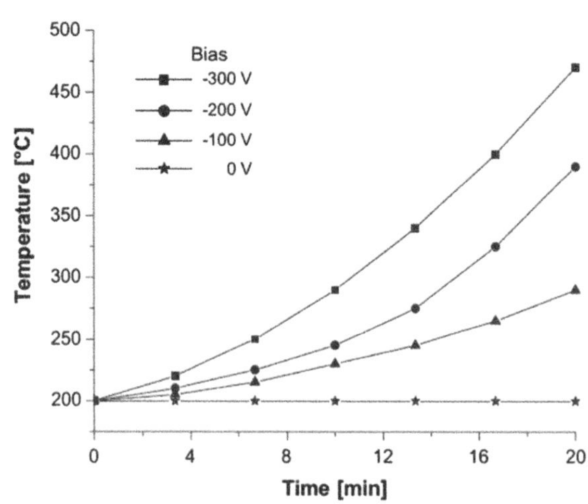

Fig. 5.1 Substrate temperature as a function of the bias voltage [9]

It was observed that for null bias, the microstructure mainly consisted of austenite, but increasing the absolute value of bias voltage and consequent temperature increment led to widening and lower 2θ shifting of austenite peaks (formation of expanded austenite). For example, in −300 V, the substrate temperature reached 470 °C and the most intensity of γ_N was obtained.

The nitrided depth also depended on bias voltage and temperature. For null bias, almost no layer was formed. However, increasing temperature led to thickening of nitrided layer. In bias voltage of −300 V, the substrate temperature reached 470 °C and the thickness of nitrided layer was 10–11 μm.

According to other research published from these authors [10], the bias voltage and consequently substrate temperature as well as nitrogen concentration had a significant effect on the shift value of γ_N.

The intensity of γ_N as well as the shift to smaller 2θ angles are seen in Fig. 5.2. As seen, it depends on the sample temperature and the nitrogen density within the γ phase. From a null bias (200 °C) to the bias of −300 V (480 °C), the γ_N phase shifted to lower 2θ angles and its intensity increased. After increasing bias voltages beyond −400 V (550 °C), the diffraction pattern changed dramatically and new peaks appeared. This could be due to the transformation of γ_N into ferritic phases and CrN.

Negm [11] investigated the effect of post-annealing on the properties of plasma nitrided AISI 304 austenitic stainless steel. The applied RF plasma nitriding parameters were frequency of 13.56 MHz, atmospheres of 50 % H_2 + 50 % N_2 and 75 % H_2 + 25 % N_2 and 100 % N_2, and duration of 10 min. The samples were annealed at temperatures of 200, 300, 400, 500, 600, 700, 800, and 900 °C for 1 h.

The compound layer thickness was almost constant up to 400 °C. After that, the thickness began to increase. The compound layer thickness for sample treated using 75 % H_2 + 25 % N_2 was far smaller with respect to thickness of the other two samples. Moreover, it remained constant for all nitriding temperatures.

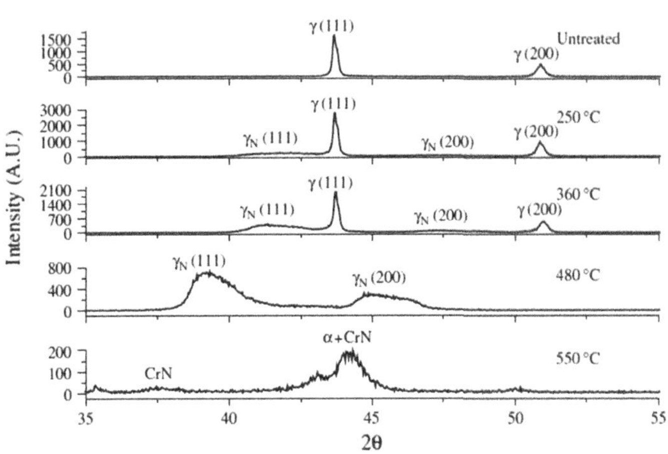

Fig. 5.2 X-ray diffraction spectra of AISI-304 at different temperatures [10]

Nitriding using 50 % H_2 + 50 % N_2 led to a decrease in grain size and increase in the thickness of grain boundaries but annealing smeared out most of the grain boundaries. The formation of distinct nitrided layers and thickening of compound layer were observed with post-annealing at 800 °C. Similar effects were reported for the atmosphere of 75 % H_2 + 25 % N_2.

As shown in Fig. 5.3, for 50 % N_2 + 50 % H_2 atmosphere, the phase composition differed with increasing annealing temperature. γ_N and γ' disappeared but α increased and evidence of CrN was detected. The fact that more α formed at the expense of γ verified the changing of sample composition, where the nitrogen is leaving the compound layer via diffusion into the bulk. This trend toward loosing nitrogen as the temperature increased to form principally α-phase continues up to 900 °C as shown by the X-ray data. In case of 25 % N_2 + 75 % H_2, there was no sign of changing phase composition up to 600 °C. Afterwards α and CrN began to appear and grow. It was reported that, annealing up to 800 °C produced more α and lower γ phases.

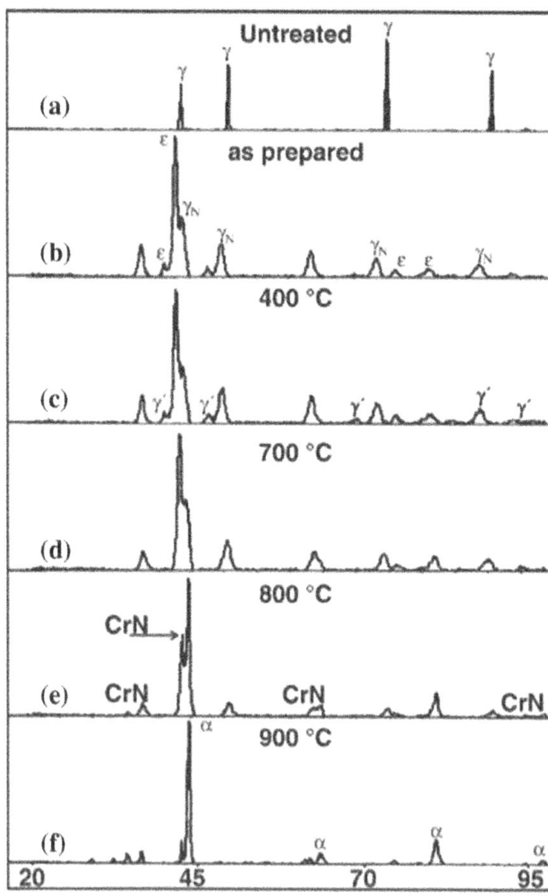

Fig. 5.3 X-ray diffraction patterns of plasma treated sample at 50 % N_2 + 50 % H_2 before and after annealing [11]

Table 5.1 Phase composition of RF plasma nitrided AISI 304 in different atmospheres [12]

Gas composition	Pure N_2	12 % H_2	35 % H_2	50 % H_2	75 % H_2
Phases	γ'-Fe_4N and a few CrN	γ'-Fe_4N, CrN and ε-$Fe_{2-3}N$	Similar to pure N_2 with more γ'-Fe_4N	More CrN, γ_N and a few γ'-Fe_4N	α, CrN and traces of γ

Negm [12] investigated the effect of different gas mixtures on the microstructure of plasma nitrided AISI 304 in the constant power of 520 W for 10 min.

The increment of sample temperature was observed with increasing $H_2/(N_2 + H_2)$ ratio up to 75 %. This showed the great importance of gas composition in plasma treatment. The obtained results for phase composition of nitrided layer in different atmospheres are reported in Table 5.1.

Kumar et al. [13] reported the effects of hydrogen amount of the nitriding atmosphere on the properties of nitrided layer. They applied these parameters: frequency of 13.56 MHz, temperature of 400 °C, duration of 3 h, and different hydrogen percentages ranging from 0 to 75 %.

The maximum thickness of the treated layer was obtained at 15 % H_2. Hydrogen concentration of 5–50 % enhanced the nitriding efficiency. At 75 % H_2, the nitrided layer was thinner than pure N_2.

Expanded austenite was present at all of the hydrogen percentages. Because of nitrogen concentration gradient in the treated layer, the expanded austenite peaks were severely broadened and showed an asymmetry. For samples treated either in pure nitrogen or with 75 % hydrogen, this effect was more obvious, implying that the samples nitrided with 5–50 % hydrogen were relatively ordered and exhibited a smaller nitrogen concentration gradient. For 75 % hydrogen, the lattice parameter a for the expanded austenite was calculated to be 0.375 nm and for the other treatment regimes 0.381 nm, compared with 0.361 nm for the substrate phase of austenite. This was probably due to the lower concentration of nitrogen in the treated layer. The amount of the γ phase was reversely correlated with the thickness of the expanded austenite layer. This was due to the less detection of substrate austenitic phase with thickening of expanded austenite layer. They also reported the improved coherence in the expanded austenite structure caused by addition of hydrogen.

El-Hossary [4] studied the influence of surface microcracks and temperature gradients on the RF plasma nitriding rate of AISI 304. Two groups of samples were treated and studied: (1) constant power of 520 W with durations of 3, 4, 5, 6, 7, 10, 15, 18, 25, and 30 min and (2) constant time of 10 min with powers of 370, 400, 425, 450, 475, 500, 550, 600, 650, and 750 W.

For the fixed power, duration of 7 min resulted in the highest nitriding rate (3 $\mu m^2/s$). For the least time (3 min) the nitriding rate was 0.3 $\mu m^2/s$. This value was still very high compared to other methods of nitriding. For fixed time, the highest nitriding rate of 2.67 $\mu m^2/s$ was achieved for the sample that was treated with the power of 500 W.

A linear relationship between the surface temperature and the input plasma power was achieved. When the plasma power was close to 375 W, the surface temperature was 380 °C. By increasing plasma power to 750 W, the surface temperature increased to 650 °C.

According to Fig. 5.4, the thickness and concentration of nitride locations increased with increasing applied power (up to 500 W). However, with a further increase in the applied plasma power, the thickness of the layer decreased.

By increasing the nitriding time up to 7 min, the thickness and concentration of nitride locations increased. However, beyond 7 min, a change was observed in this trend.

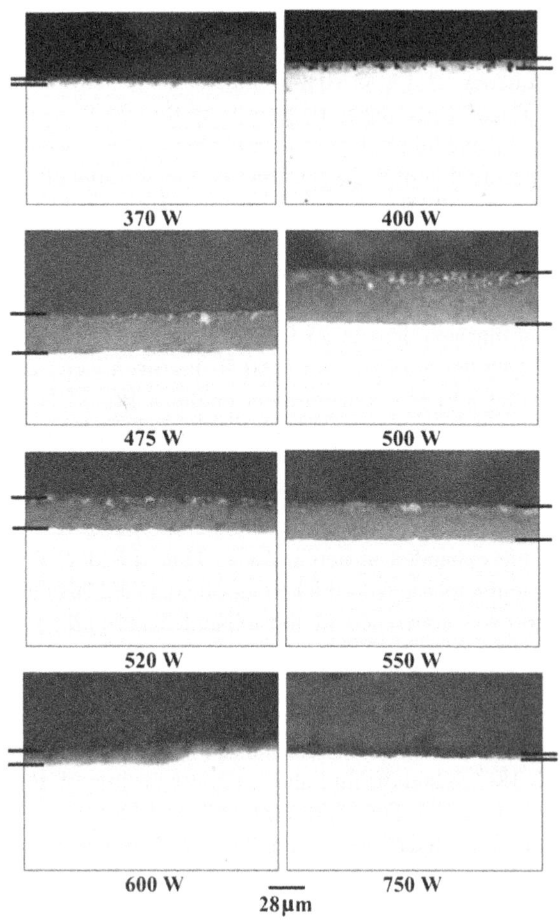

Fig. 5.4 Microstructure cross section of compound layer treated for 10 min at different plasma powers: 370, 400, 475, 500, 520, 600, and 750 W [4]

Up to 7 min, the sample temperature was between 500–600 °C and the sample surface was full of microcracks in this state. With these conditions, the active nitrogen ion species diffused into these microcracks and nitriding effect increased. However, when the time exceeded 7 min, the layer thickness decreased with increasing time. In the case of different powers, above 500 W, the temperature of the sample increased to reach more than 600 °C and the surface was almost free of cracks.

de la Piedad-Beneitez et al. [14] studied a 15 % N_2 + 85 % H_2 inductive RF plasma applied to the PIII[1] nitriding of AISI 304 steel.

The biasing and ion bombardment increased the sample temperature from 350 to 500 °C. This can affect the compound layer thickness and intensity of expanded austenite. The detected nitrides were Fe_3NiN and FeNiN. Temperature increment resulted in the enhancement of nitrogen diffusivity and further broadening of γ_N peaks.

5.1.2 Martensitic Stainless Steel

Kim et al. [15] studied RF plasma nitriding of AISI 420 martensitic stainless steel. Four treatments used the temperature of 400 °C, with the varying duration in the range 1–16 h, and four treatments used a process time of 4 h, with temperatures in the range 350–500 °C. The applied frequency was 13.56 MHz.

Figure 5.5 shows XRD patterns of the samples. Supersaturation of martensite with nitrogen corresponds for peaks' shifting phenomena seen in Fig. 5.5. Although, martensite is a supersaturated structure itself, but this could be the best label for explaining the shifting phenomena as described in the case of austenitic stainless steel. It could be indicated that the nitrogen-rich layer was thick enough, even at low temperature or short treatment time, that the peak from the underlying martensitic substrate was not visible.

Transformation depends on the combination of temperature and treatment time. Because of this, in the case of 350 °C treated sample (Fig. 5.5a), the peak at 38.5° can belong to CrN phase, although this temperature was low for transformation to CrN. However, with differing chemical composition, it could be possible. For varying time samples (Fig. 5.5b) a transformation to CrN and ferrite in the samples treated for 9.0 and 16.0 h is observable. The formation of CrN in such a temperature had a deleterious effect on properties of treated layer, so that 4 h and 400 °C were close to the optimum treatment conditions. Under these conditions, the treated layer retained its martensitic structure, but with a larger lattice parameter than the bulk.

[1] Plasma Immersion Ion Implantation.

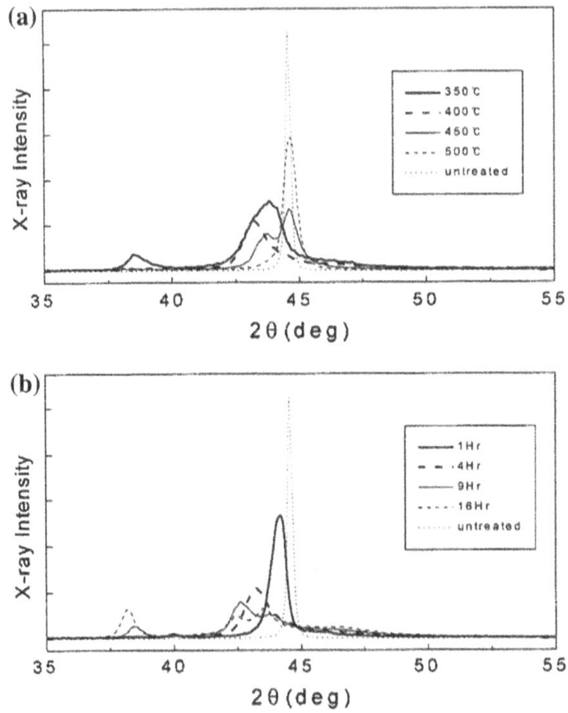

Fig. 5.5 Sections of XRD scans for samples of AISI 420 martensitic stainless steel nitrided for **a** 4 h at the indicated temperatures and **b** 400 °C for the indicated durations. Results from an untreated sample are included in each panel [15]

5.1.3 *Tool Steels*

Akbari et al. [16] studied the effects of microstructure on the plasma nitriding behavior of AISI M2 high-speed steel. Before nitriding, they did two kinds of heat treatment on the samples according to Fig. 5.6. Samples of the group A were

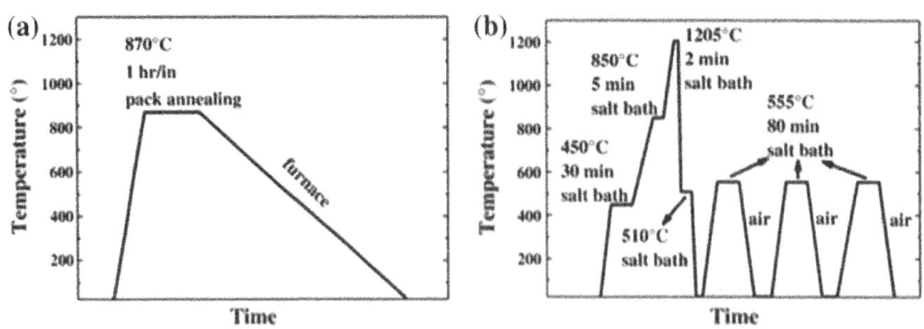

Fig. 5.6 Applied time/temperature-sequence diagrams for heat treatment of AISI M2 high speed tool steel, **a** full annealing and **b** quenching–tempering [16]

subjected to the full annealing treatment and samples of the group T were subjected to the quench and triple tempering treatment. The samples were plasma nitrided in the atmosphere of 25 % N_2 + 75 % H_2, under pressure of 7.5 Pa, at temperature of 450 °C for 8 h and with frequency of 13.56 MHz.

Before nitriding, the microstructure of group A samples consisted of alloy carbides in the ferritic matrix and the hardness was 230 HV_{30}. For samples of group T, dispersed carbides in a martensitic structure were observed with hardness of 870 HV_{30}.

After plasma nitriding, for all of the samples, the diffusion zone was visible but the white layer was not observed. The diffusion zone for samples of group A was thicker because of the presence of iron nitrides γ' and ε in the nitrided A samples. According to the Fe–N phase diagram γ' and ε phases may form if the nitrogen concentration reaches a certain level at the sample surface. This means the deeper diffusion of nitrogen. As seen in Fig. 5.7, the main difference between unnitrided A and T samples was microstructure. During annealing, due to the lower cooling rate, the temperature of the sample remained above the pearlitic nose for a longer duration and this provided a sufficient time for carbon and other alloying elements to diffuse out from iron lattice. However in the case of quenching, the very high cooling rate resulted in the remaining of carbon and alloying elements inside the BCT lattice. These reasons can be mentioned for the reduction of nitrided depth:

1. The reduction of the nitrogen diffusion coefficient in the BCT-α lattice due to the formation of solid solution and trapping. Carbon as an interstitial alloying element reduced more effectively the number of available interstitial sites for nitrogen. The other important mechanism was trapping. During trapping, nitrogen atom was temporarily immobilized in Fe–M–N configurations, without the direct formation of nitride precipitates.
2. Formation of nitride precipitates in the diffusion zone. M2 steel had a high concentration of strong and moderate nitride forming elements and consequently can form simple and complex nitrides precipitates.
3. Compressive residual stresses associated with the martensite structure which should modify and reduce nitrogen diffusion toward the sample core.

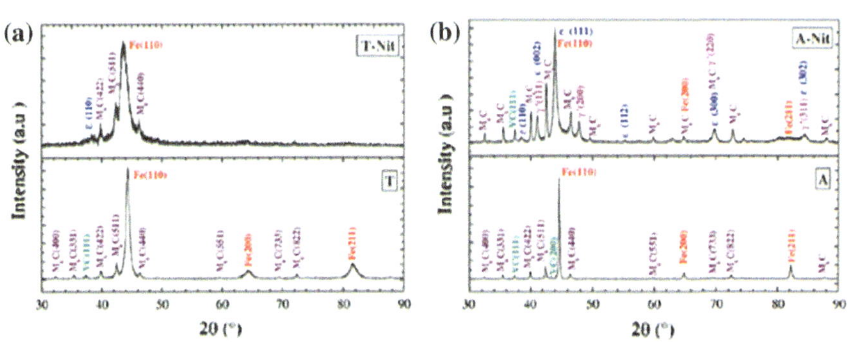

Fig. 5.7 X-ray diffractograms for unnitrided and nitrided substrates in the **a** hardened–tempered and **b** fully annealed M2 steel [16]

The microstructure of nitrided T samples consisted of a few amount of ε nitride. However, for A samples γ'-Fe_4N was also observed. Some other main peaks belong to substrate were also observed as seen in Fig. 5.7.

Akbari et al. [5] investigated the effect of different nitriding atmospheres on the properties of the RF nitrided samples. The applied parameters were: temperature of 450 °C, pressure of 7.5 Pa, and frequency of 13.56 MHz. The different applied gas mixtures were as follows: 25 % N_2 + 75 % H_2, 50 % N_2 + 50 % H_2 and 75 % N_2 + 25 % H_2.

The thickness of diffusion zone (in fact nitrided layer) was changed with gas composition. A great increase was observed for the sample nitrided with 50 % N_2 compared with the sample nitrided with 25 % N_2. However, no noticeable increment was observed for atmosphere of 75 % N_2. The case depth of nitrided samples was estimated to be 70, 90, and 105 μm for the 25 % N_2, 50 % N_2 and 75 % N_2.

In the case of 25 % N_2, almost no iron nitrides were formed indicating that no compound layer was formed in this condition. However, the supersaturation of matrix phase was obvious. On the other hand, applying the atmosphere of 50 % N_2 and 75 % N_2 resulted in the formation of the monophase ε-$Fe_{2-3}N$ compound layer. This was owing to the fact that ε-$Fe_{2-3}N$ phase was formed when the carbon content is high contrary to γ'-Fe_4N that needs less carbon to be formed. In the conventional DC plasma nitriding, high primary sputtering rate led to the decarburization of the surface [17] and consequently lowering the carbon content and formation of γ'-Fe_4N. In RF plasma nitriding the energy of the ions arriving on the surface was not enough to sputter the surface and the formation of ε-$Fe_{2-3}N$ was favored.

5.1.4 Constructional and Structural Steels

Keddam et al. [7] studied the nitrided layer of XC38 carbon steel obtained by RF plasma nitriding. The nitriding atmosphere was 60 % N_2 + 40 % H_2. The other parameters are presented in Table 5.2 as well as compound layer thicknesses obtained.

Table 5.2 shows that with increasing temperature, the compound layer thickness (thickness of γ' layer) increased according to the Eq. 5.1:

$$x = \alpha \times t^n, \tag{5.1}$$

Where α and n are material constants depending on the nitriding conditions. For these nitriding conditions the constants were estimated as $\alpha = 2.47$ and $n = 0.288$ showing the nonuniformity of the compound layer thickness. This deviation from a classical parabolic growth could be because of preferential nucleation and growth of the γ'-Fe_4N phase along grain boundaries. Some energetically preferred places such as grain boundaries and structural defects are suitable places for the nucleation starting. At some points along the interface, the compound layer extended into the grains boundaries. The formation of this iron nitride was then affected by internal

Table 5.2 The change of the compound layer thickness with RF plasma parameters [7]

Process parameters		Compound layer cthickness (μm)
Temperature (°C)	Time (h)	
350	4	0–0.77
400	4	
500	3	1.1–3.3
	4	3.5–4.5
	8	3.8–4.9
	16	4–6.5
570	4	4.6–5.4

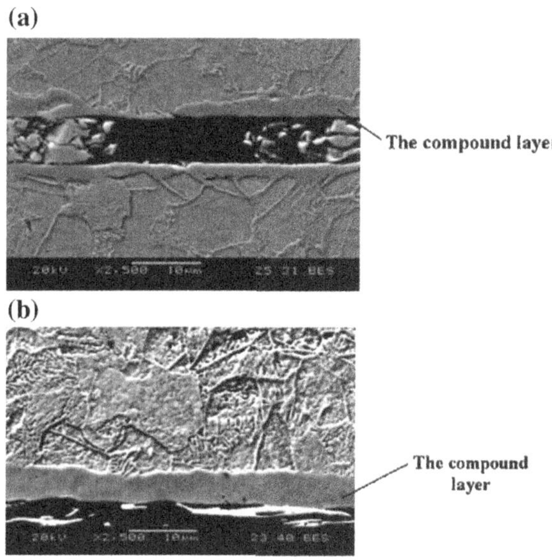

Fig. 5.8 SEM photomicrographs; **a** nitrided 3 h at 500 °C, **b** nitrided 4 h at 570 °C [7]

stresses produced by the distortion of the lattice due to dissolved nitrogen, where the nitrogen diffusion is promoted by the presence of H_2 in the gas mixture.

It was reported that the compound layer of all of the samples consisted of γ'-Fe_4N.

According to Fig. 5.8, it is clear that the compound layer had a nonuniform thickness for a treatment of 500 °C for 3 h, whereas its thickness was regular and continuous for a temperature of 570 °C and duration of 4 h. Besides, the micropores were not observed for the latter.

Bouanis et al. [18] studied microstructural behavior of RF plasma nitrided C38 carbon steel. They applied these parameters: pressure of 16.8 Pa, frequency of 13.56 MHz, atmosphere of N_2, and durations of 6, 8, and 10 h.

The microstructure of untreated sample consisted of ferrite and pearlite with grain size of 20–50 μm.

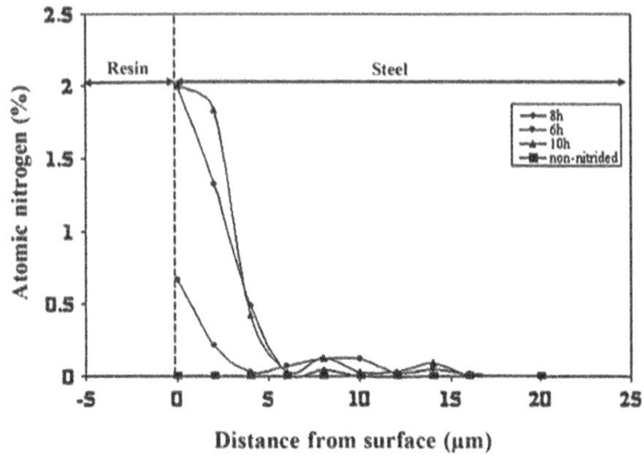

Fig. 5.9 Quantification of nitrogen along a line from the inside of the sample to its edge for untreated and treated in pure nitrogen for different times of plasma treatment [18]

For duration of 6 h, no significant nitrided layer was obtained. For 8 h nitrided sample, a compound layer (with thickness of 10 μm) and a nitrogen diffusion zone were formed. And for 10 h, the thickness of nitrided layer increased to 14 μm.

As can be seen from Fig. 5.9, for all treated samples, the highest concentration of nitrogen was that of the edge. Nitrogen profiles showed that from the edge to the inside, nitrogen content decreased. This decrease was faster for the sample treated for 6 h in comparison to the other samples. The concentration at the edge was 2 % of nitrogen for the 10 h nitrided samples, the same to that of 8 h plasma treated. However, the nitrogen content remained more constant in the compound layer for the 10 h nitrided samples, whereas it decreased slightly for 8 h of treatment time. Consequently, the thickness and nitrogen content mainly depended on the nitriding time.

5.2 Mechanical Properties

Valencia-Alvarado et al. [9] (see Sect. 5.1.1) concluded that increasing the nitriding temperature led to higher hardness values. As shown in Fig. 5.10, the increase in temperature clearly seemed to favor the ion penetration which led in turn to the formation of deeper nitrided layers. As noted by the authors, because of forming no chromium nitrides, the high degree of improved hardness was due to a considerable concentration of nitrogen in the γ austenitic phase.

Negm [11] (see Sect. 5.1.1) investigated the relation of post-annealing temperature with surface microhardness. It was reported that the post-annealing with temperature up to 400 °C resulted in nearly stable microhardness values. Further annealing, led to a gradual decrease in the microhardness values of the samples to reach 1200 HV for samples that were treated using 100 % N_2 or 50 % H_2 + 50 % N_2 and annealed at 800 °C. At the end, all the samples had greater microhardness values relative to the associated values of untreated sample.

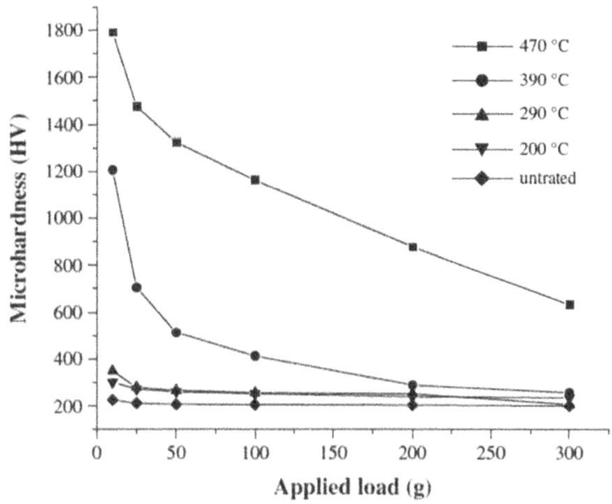

Fig. 5.10 Microhardness versus applied load [9]

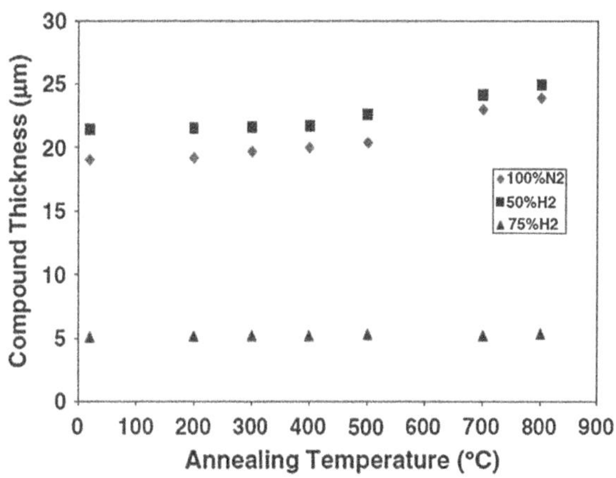

Fig. 5.11 Thickness variations of compound layer with annealing temperatures for plasma treated samples using different H_2/N_2 ratios [11]

Figure 5.11 shows the variation in the thickness of the compound layer with annealing temperatures obtained for different hydrogen percentage content. For samples annealed with temperatures up to 400 °C, the thickness of the compound layers was almost constant. After that, the thickness increased with annealing temperature up to 800 °C. As seen in figure, the behavior of the compound layers for samples treated using 100 % N_2 and 50 % H_2 + 50 % N_2 were nearly similar. It can be also seen that for each annealing temperature, the thickness values were nearly constant.

de la Piedad-Beneitez et al. [14] (see Sect. 5.1.1) studied the effects of different temperatures and pressures on the microhardness of the nitrided samples. They observed that at higher temperatures the deeper nitrided layers were obtained due to the greater ion penetration. The hardness increment with temperature could be explained by ion bombardment of the surface by hydrogen and nitrogen.

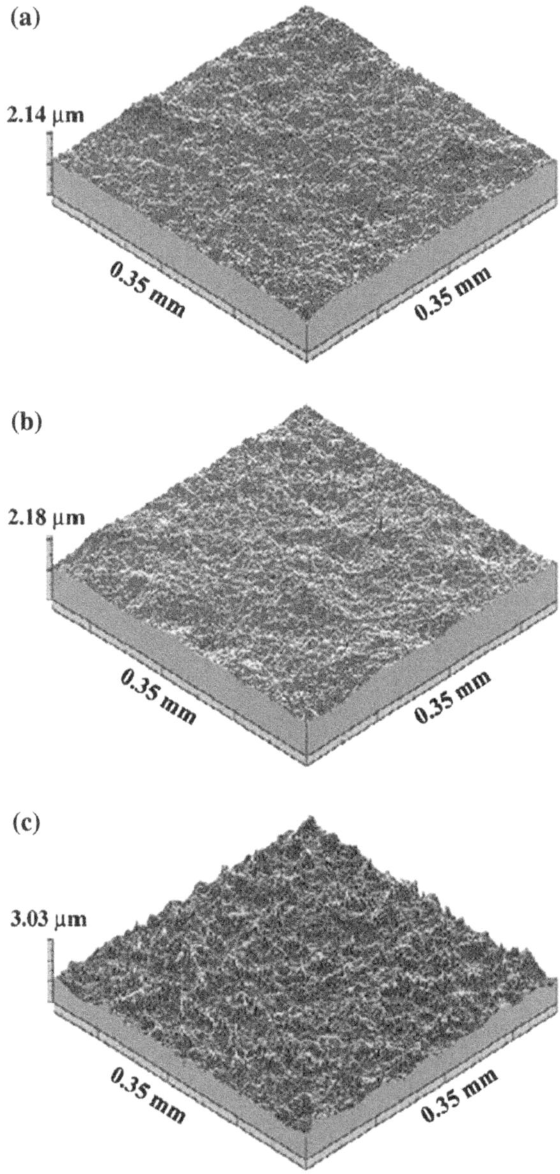

Fig. 5.12 3D surface profilometry of M2 steel **a** in the unnitrided state (representative of A and T samples) and **b** in the nitrided T samples and **c** nitrided A samples [16]

Akbari et al. [16] (see Sect. 5.1.3) studied the effects of initial microstructure and post-plasma nitriding on the surface roughness of AISI M2 tool steel. Figure 5.12 shows the surface profilometry of samples in different surface states. The roughness data calculated shows that plasma nitriding increased the surface roughness of the substrate and that this increase was twice greater for annealed and nitrided samples.

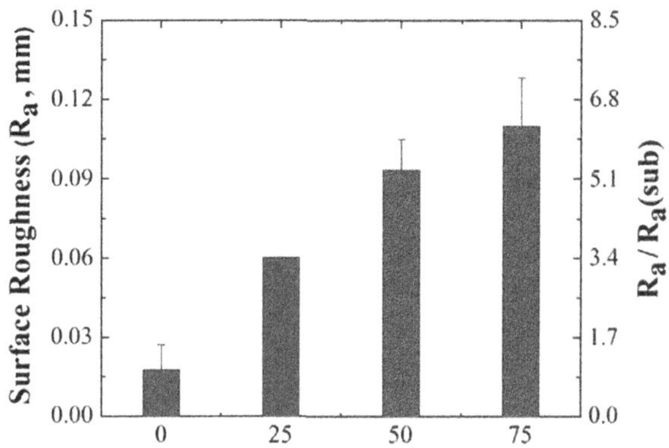

Fig. 5.13 Surface roughness (R_a, root mean square) results of AISI M2 tool steel before and after RF plasma nitriding at different gas compositions [5]

Akbari et al. [5] (see Sect. 5.1.3) studied the effects of different nitriding atmospheres on the roughness and wear properties of AISI M2 tool steel. As seen in Fig. 5.13, the least and the most surface roughnesses belonged to the samples nitrided under 25 and 75 % N_2 atmosphere, respectively. "The increase in surface roughness during ion nitriding may be caused by contribution of three distinct phenomena: (i) surface roughening in part by different lattice expansion of grains of polycrystalline material with the inward nitrogen diffusion from the differential swelling of the grains and in part by volume difference between the alloy nitrides inside the diffusion zone and the matrix, (ii) sputtering caused by the collision of energetic particles accelerated from plasma toward the surface, (iii) redeposition of the sputtered material on the surface. As the nitriding process performed in a floating potential, the energy of ions arriving to the surface was only about 15 eV. Such low energies cannot generate sputtering; therefore, the increase of surface roughness after nitriding can be entirely attributed to the volumetric lattice expansion associated to the nitrogen diffusion. Thus, inward diffusion of the higher nitrogen contents results in higher surface roughness".

The best and the worst wear resistance were resulted by 25 and 75 % N_2 atmosphere, respectively. This was ascribed to higher surface hardness, higher supersaturation degree (and consequently higher residual compressive stress), and no formation of compound layer on the surface of the sample nitrided using 25 % N_2. As noted before, in the case of 25 % N_2 almost no compound layer was formed but increasing the nitrogen amount led to a monophase compound layer consisted mainly of ε-$Fe_{2-3}N$. This phase is inherently a brittle phase and under a light load began to be cracked. Therefore, the formation of this phase had a deleterious effect on the wear properties of nitrided sample.

Bouanis et al. [18] (see Sect. 5.1.4) studied the behavior of microhardness of C38 steel after plasma nitriding. It was concluded that the surface hardness of nitrided samples increased with time. The maximum hardness value they obtained was 1487 $HV_{0.005}$ for duration of 8 h. This trend resulted in the increase of nitrogen concentration and this can lead to the formation of a thicker nitrided layer on the steel surface and consequently better hardness values. They also observed the

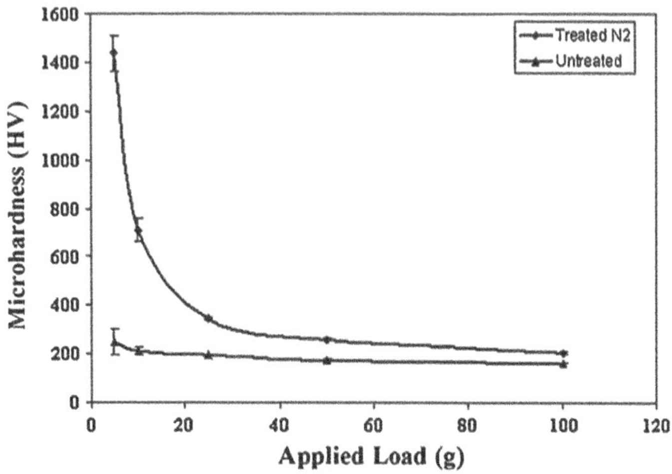

Fig. 5.14 Variation of the Vickers microhardness of the untreated and 8 h nitrided C38 steel as function of the applied loads [18]

behavior of surface hardness values with different applied loads for measuring hardness. It can be observed in Fig. 5.14 that except for the untreated sample, the microhardness values rapidly decrease with increasing applied load.

5.3 Corrosion Behavior

Ibrahim et al. [8] (see Sect. 5.1.1) studied corrosion behavior of annealed and as-treated AISI 304 samples plasma nitrided with immersion in 5 % HCl solution. Figure 5.15 shows the behavior of microhardness and weight loss of different

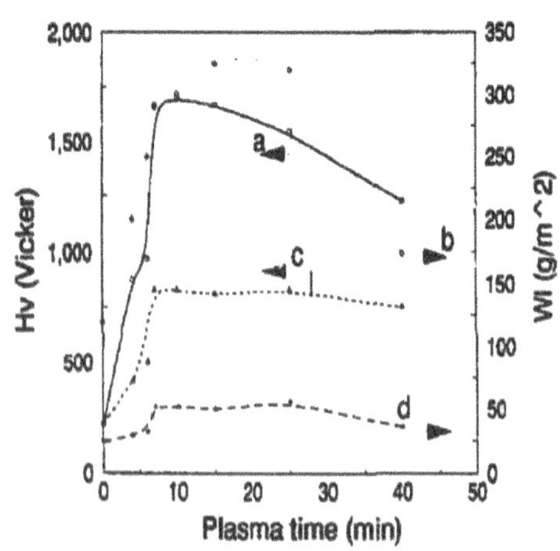

Fig. 5.15 Microhardness and weight loss per square meter of austenitic stainless steel and nitrided austenitic stainless steel versus plasma time for as prepared (*a* and *b*) and annealed (*c* and *d*) samples [8]

samples. As can be seen, up to 7 min of plasma nitriding, the values increased sharply and both their trends were similar. After a steady state, the values dropped to lower values but again with a similar trend. Also it can be indicated that after annealing, they both kept their trends but with lower values.

Figures 5.16 and 5.17 show the weight loss and corrosion rate of untreated and treated samples before and after annealing. It is obvious that annealing decreased both weight loss and corrosion rate. In both figures, three distinct areas or stages are visible. First, when the immersion time was few hours, the weight loss rate was high. Thereafter in the second stage, the lowest weight loss rate was obtained. Finally in the third stage, after the end of the second stage, the highest weight loss rate can be seen. For annealed samples (treated and untreated), with increasing the immersion times, the corrosion rates decreased exponentially to low values.

To evaluate the effect of different exposure times and nitriding durations, they studied the corrosion behavior of samples in different durations. They observed that for all samples, the weight loss rate increased with increase in immersion time. Furthermore, the lowest weight loss rate belonged to the untreated and followed by

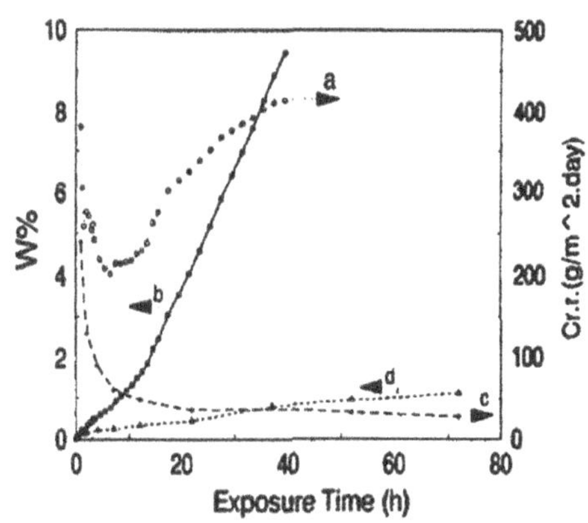

Fig. 5.16 Weight loss % and corrosion rates of untreated samples versus exposure time for not annealed (a and b) and annealed (c and d) samples [8]

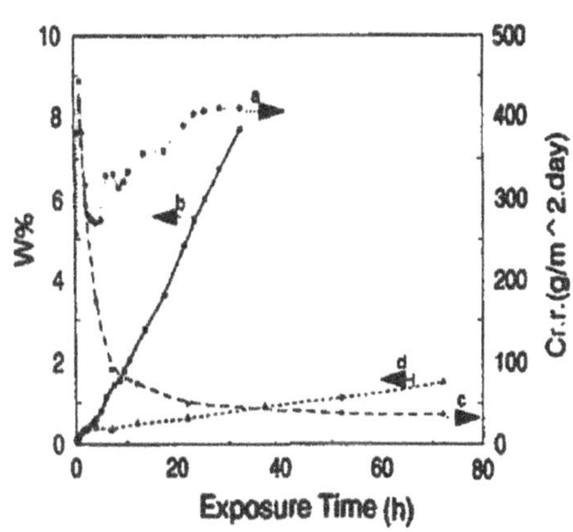

Fig. 5.17 Weight loss % and corrosion rates of nitrided samples treated for 40 min versus exposure time for not annealed (*a* and *b*) and annealed (*c* and *d*) samples [8]

samples which were nitrided for a short time up to 15 min. Thereafter increasing nitriding time led to a decrease in the weight loss values. For annealed samples this behavior was also reported although annealing led to a decrease in the weight loss values. In this study, the improvement of corrosion behavior after annealing was proved because the corrosion rates of all annealed samples were less than those of the as-prepared ones.

The dissolution of chromium nitride and/or carbide precipitation in the matrix of plasma nitrided samples occurred as a result of annealing at 800 °C. The following rapid cooling avoided the reprecipitation in the nitrided and annealed samples. This can reduce the hardness of annealed samples with improving the corrosion resistance. The samples that were not annealed might contain different phases such as ε, τ, τ′, CrN, and α. This heterogeneous phase structure had deleterious effect on the corrosion resistance. Contrarily, the annealed samples, contained mainly τ, α, and small traces of CrN with a homogenous structure. Therefore, it can be concluded that annealing treatment of nitrided samples reduced the precipitation of nitride and/or carbide, dislocation density and localized stress, and increased the grain sizes and homogeneity of structural phases. All of these led to a decrease in the hardness of the samples related to the samples which were not annealed. Consequently, the corrosion resistance of nitrided-annealed samples was much better than those of just nitrided samples.

Bouanis et al. [19] widely studied the corrosion behavior of plasma nitrided C38 steel. They used gravimetric tests based on weight loss rate with immersion in 1 M HCl solution for 24 h. Besides, they did potentiodynamic tests in the same solution. Two parameters were calculated: corrosion rate and inhibition efficiency using Eqs. 5.2 and 5.3.

$$\vartheta_{\mathrm{corr}}(\mathrm{mpy}) = \frac{3.45 \times 10^{6} W}{ADT}, \tag{5.2}$$

Where W is the weight loss (g), D the density of the C38 steel (7.85 g cm^{-3}), A the area of the coupon (cm^2) and T is the exposure time (h). With calculating and using the values of weight loss of untreated (W_0) and plasma nitrided steel (W), the inhibition efficiency ($E\%$) was determined using the equation below

$$E(\%) = (1 - \frac{W}{W_0})100 \tag{5.3}$$

After doing immersion test, all of the data obtained were presented in Table 5.3. It is obvious that plasma nitriding significantly decreased the weight loss. Also, increasing the nitriding duration led to even lower weight loss. The maximum inhibition efficiency was obtained for 8 h nitriding (97 %) and further increment of time did not have any significant effect on the corrosion rate and weight loss values.

The results of potentiodynamic tests for different nitriding times are shown in Table 5.4 and Fig. 5.18. As shown, the best result belonged to 8 h nitriding.

Table 5.3 Corrosion parameters obtained from weight loss measurements for untreated and plasma nitrided C38 steel after 24 h immersion period in 1 M HCl at 30 °C [19]

Sample	Time of plasma treatment (h)	*W* (g)	υ_{corr} (10^3 mpy)	*E* (%)
Untreated sample	0	0.78192	3.182	–
Treated sample	1	0.41480	1.688	81.9
	2	0.12636	0.514	83.8
	3	0.03197	0.130	95.9
	8	0.02398	0.097	96.9

Table 5.4 Polarization parameters for untreated and plasma nitrided C38 steel after immersion in 1 M HCl at 30 °C [19]

Sample	Time of plasma treatment	E_{corr} versus SCE (mV)	I_{corr} (mA cm^{-2})	b_c (mV dec^{-1})	*E* (%)
Untreated	0	−470	554	186	–
Treated	1	−481	97	166	82.5
	2	−475	74	165	86.6
	3	−468	65	164	88.3
	8	−450	56	163	89.9

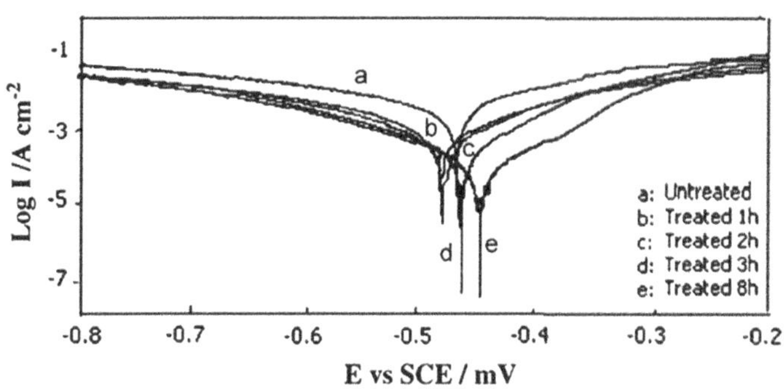

Fig. 5.18 Polarization curves of untreated and plasma nitrided C38 steel after immersion in 1 M HCl at 30 °C [19]

The reduction of I_{corr} with nitriding duration is obvious but for E_{corr} no definite trend was observed. The cathodic Tafel slopes (b_c) were also slightly improved by plasma nitriding. These results suggested that the plasma treatment reduced anodic dissolution and retarded the hydrogen evolution reaction. Inhibition efficiency was also calculated for Tafel polarization study by means of following equation:

$$E(\%) = \frac{I^0_{corr} - I_{corr}}{I^0_{corr}} 100 \tag{5.4}$$

in which I^0_{corr} and I_{corr} are the corrosion current density values of untreated and plasma nitrided steel, respectively. As seen in Table 5.4, the results of immersion tests were proved by polarization test.

According to XPS studies, they concluded that the important factor in improved corrosion resistance of nitrided samples was production of Fe^{3+} ions and a dense, stable, and insoluble oxide layer (FeOOH or γ'- or ε-Fe_2O_3) to decrease the diffusion of ions. All of these results were confirmed by Bouanis et al. [20].

References

1. Priest J M, Baldwin M J, Fewell M P, Haydon S C, Collins G A, Short K T, Tendys J (1999) Low pressure r.f. nitriding of austenitic stainless steel in an industrial-style heat-treatment furnace. Thin Solid Films 345:113-118
2. Liang W, Xiaolei X, Bin X, Zhiwei Y, Zukun H (2000) Low temperature nitriding and carburizing of AISI304 stainless steel by a low pressure plasma arc source. Surf Coat Technol 131:563-567
3. Fewell M P, Priest J M, Baldwin M J, Collins G A, Short K T (2000) Nitriding at low temperature. Surf Coat Technol 131:284-290
4. El-Hossary F M (2002) The influence of surface microcracks and temperature gradients on the rf plasma nitriding rate. Surf Coat Technol 150:277–281
5. Mohammadzadeh R, Akbari A, Drouet M (2014) Microstructure and wear properties of AISI M2 tool steel on RF plasma nitriding at different N_2–H_2 gas compositions. Surf Coat Technol 258:566-573
6. Baldwin M J, Fewell M P, Haydon S C, Kumar S, Collins G A, Short K T, Tendys J (1998) Rf-plasma nitriding of stainless steel. Surf Coat Technol 98:1187-1191
7. Keddam M (2008) Characterization of the nitrided layers of XC38 carbon steel obtained by R. F. plasma nitriding. Appl Surf Sci 254:2276–2280
8. Ibrahim M M, El-Hossary F M, Negm N Z, Abed M (1992) Effect of RF plasma nitriding time on microhardness and corrosion resistance of 304 stainless steel. Appl Surf Sci 59:253-260
9. Valencia-Alvarado R, de la Piedad-Beneitez A, de la Rosa-Va´zquez J, Lo´pez-Callejas R, Barocio S R, Godoy-Cabrera O G, Mercado-Cabrera A, Pen˜a-Eguiluz R, Mun˜oz-Castro A E (2008) Nitriding of AISI 304 stainless steel in a 85 % H_2/15 % N_2 mixture with an inductively coupled plasma source. Vac 82:1360–1363
10. Valencia-Alvarado R, de la Piedad-Beneitez A, de la Rosa-Va´zquez J, Lo´pez-Callejas R, Barocio S R, Godoy-Cabrera O G, Mercado-Cabrera A, Pen˜a-Eguiluz R, Mun˜oz-Castro A E (2007) γ_N-shift as a function of N_2 content in AISI 304 nitriding. Vac 81:1434–1438
11. Negm N Z (2006) Effect of annealing temperature on properties of H_2/N_2 rf plasma-treated stainless steel. Surf Coat Technol 201:1763–1767
12. Negm N Z (2006) A study on rf plasma nitriding at a constant power in different H_2–N_2 mixtures at different temperatures. Mater Sci Eng B 129:207–210
13. Kumar S, Baldwin M J, Fewell M P, Haydon S C, Short K T, Collins G A, Tendys J (2000) The effect of hydrogen on the growth of the nitrided layer in r.f.-plasma-nitrided austenitic stainless steel AISI 316. Surf Coat Technol 123:29–35
14. de la Piedad-Beneitez A, Valencia-Alvarado R, López-Callejas R, Rojas-Olmedo I A, Peña-Eguiluz R, Mercado-Cabrera A, Barocio S R, Muñoz-Castro A E, Rodríguez-Méndez B G (2011) Optimized AISI 304 steel nitriding in inductive RF N_2-H_2 plasmas. Vac 85:1149-1151

15. Kim S K, Yoo J S, Priest J M, Fewell M P (2003) Characteristics of martensitic stainless steel nitrided in a low-pressure RF plasma. Surf Coat Technol 163–164:380–385
16. Akbari A, Mohammadzadeh R, Templier C, Riviere J P (2010) Effect of the initial microstructure on the plasma nitriding behavior of AISI M2 high speed steel. Surf Coat Technol 204:4114–4120
17. Egert P, Maliska A M, Silva H R T, Speller C V (1999) Decarburization during plasma nitriding. Surf Coat Technol 221:33–38
18. Bouanis F Z, Bentiss F, Bellayer S, Vogt J B, Jama C (2011) Radiofrequency cold plasma nitrided carbon steel: Microstructural and micromechanical characterizations. Mater Chem Phys 127:329–334
19. Bouanis F Z, Bentiss F, Traisnel M, Jama C (2009) Enhanced corrosion resistance properties of radiofrequency cold plasma nitrided carbon steel: Gravimetric and electrochemical results. Electrochem Acta 54:2371–2378
20. Bouanis F Z, Jama C, Traisnel M, Bentiss F (2010) Study of corrosion resistance properties of nitrided carbon steel using radiofrequency N_2/H_2 cold plasma process. Corros Sci 52:3180–3190

Chapter 6
Other Methods

In this section, some of the works related to other plasma nitriding methods are presented. These methods are not industrial yet and works are being done on them to optimize the working condition and efficiency.

6.1 Microwave Plasma Nitriding

Camps et al. [1] studied the microwave plasma nitriding of AISI 316 austenitic stainless steel. The treatment parameters were: frequency of 2.45 GHz, pressure of 4×10^{-4}, different H_2/N_2 ratios (with N_2 ranging from 10 to 100 %), temperature of 450 °C, and duration of 50 min.

With microhardness measurements, it was shown that the highest surface hardness (1000 ± 150 HV for the 10 g load) belonged to the sample that was treated in the atmosphere of H_2/N_2:60/40. This value showed an increment of 380 % related to the initial hardness for untreated sample. A decrease in surface hardness was also shown with increasing load, because of low thickness of the modified layer. By means of nuclear reaction analysis, the nitrogen depth penetration was determined in different atmospheres. The above-mentioned gas mixture displayed the highest depth (1.6 μm).

Figure 6.1 shows the X-ray diffraction patterns for samples nitrided under varying gas mixtures. All of them consisted of matrix austenite γ phase and expanded austenite γ_N phase. It can be seen that no CrN phase was formed during nitriding and this showed the correct selecting of temperature. Also, it is obvious that the broadening of γ_N peaks increased with nitrogen content and this was in consistence with increment of nitrogen content in the surface from 26.8 to 35.7 %. The lattice parameters were also calculated and they showed an increment too (from 3.87 to 3.93 Å).

H. Aghajani and S. Behrangi, *Plasma Nitriding of Steels*, Topics in Mining, Metallurgy and Materials Engineering, DOI 10.1007/978-3-319-43068-3_6

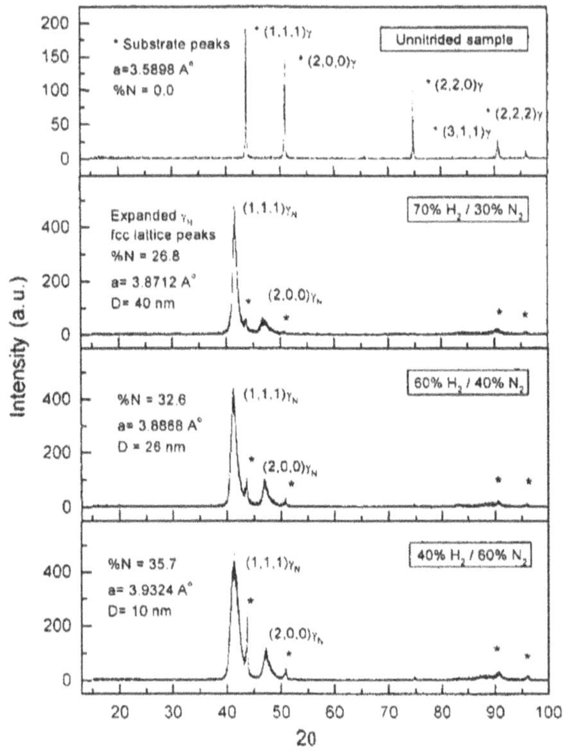

Fig. 6.1 Comparison of X-ray diffraction patterns of an untreated sample with samples treated with three different gas mixtures used to form the plasma [1]

The authors also examined the surface roughness with nitrogen content in the nitriding atmosphere. The sample prepared at H_2/N_2:70/30 had the roughness of 44 Å due to the granular morphology of the surface. However, for H_2/N_2:60/40 the surface roughness decreased to 31 Å. This was due to the changing of surface features from globular spherical grains to rod-globular grains.

The corrosion test was also done to evaluate the effect of plasma nitriding with optimum conditions mentioned above on the corrosion behavior of AISI 316 austenitic stainless steel. The improvement of approximately 90 % was obtained in the corrosion rate. It decreased from 588.6×10^{-3} to 309.6×10^{-3} mpy. This result was also consistent with XRD results showing no formation of CrN in this sample and improving corrosion behavior.

Lin et al. [2] investigated the microwave plasma nitriding of AISI 303 austenitic stainless steel. The treatment parameters are presented in Table 6.1.

According to Fig. 6.2, the nitrogen penetration depth into the substrate increased from 3.6 to 13.6 μm as treating temperature increased from 350 to 450 °C. However, further increase in the temperature to 550 °C led to a drop in the penetration depth to 4.5 μm.

Table 6.1 Process parameters used by Lin et al. for plasma nitriding of AISI 303 [2]

Parameters	Process		
	1	2	3
Temperature (°C)	350	450	550
Pressure	1	1	1
Processing time (min)	80	80	80

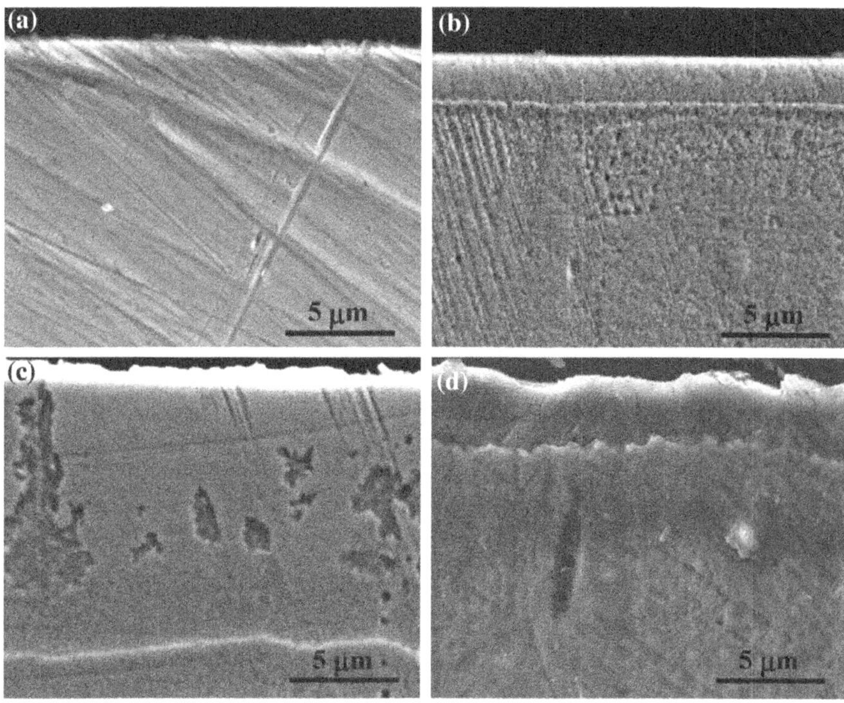

Fig. 6.2 SEM images of cross-sectional microstructure of the investigated samples: **a** untreated, treated at **b** 350 °C, **c** 450 °C, and **d** 550 °C samples [2]

Plasma nitriding led to the formation of expanded austenite instead of austenite. With increasing the temperature from 350 to 450 °C, the broadening and shifting of γ_N peaks to lower 2θ angles enhanced. With reaching the temperature to 550 °C, a microstructural transformation occurred and expanded austenite disappeared and altered with α and CrN. This was due to the decomposition of γ_N into CrN and α. The reason for this transformation was probably owing to the decomposition and chemical reaction of metastable γ_N phase with the incoming active nitrogen species during the treatment.

The microhardness measurements also displayed an increment from untreated sample to the sample nitrided at 450 °C (180 → 420 → 1250 HV), however, for 550 °C it was reduced again to 1050 HV. This phenomenon was owing to the decomposition of hard γ_N phase and formation of softer CrN and α phases.

6.2 Electron Beam Excited Plasma Nitriding

Abraha et al. [3] treated Cr–Mo low alloy steel by means of electron beam excited plasma. The parameters were: temperature of 500 °C, beam energy of 100–140 V and duration of 1.5–4.5 h.

It was observed that, the acceleration voltage plays a key role in this method of plasma nitriding. With increasing the voltage, less substrate ferritic phase was detected showing the thickening of nitrided layer with acceleration voltage. Besides, the formation of CrN, Fe_4N, and Fe_3N was enhanced with voltage. This was resulted by the enhanced dissociation of nitrogen into atomic nitrogen caused by impact of energetic electrons. Thereafter, a higher flux of nitrogen atoms and ions was produced and consequently higher interaction with the sample surface occurred. Therefore, the higher beam voltage can produce thicker layer and more nitrided compounds.

The roughness of the unnitrided sample was reported to be 9 nm. After treatment, it was changed to 420 and 32 nm for acceleration voltages of 140 and 100 V.

In the case of hardness, an ascending trend was observed with increasing the voltage from 100 to 140 V in a constant duration of 3 h. The maximum hardness of 900 HV was obtained for 140 V, while the hardness of untreated samples was 400 HV. At the depth of 60 μm from the surface, a hardness value of more than 1.5 times that of the untreated sample was observed. A similar trend was also observed for treatment duration at a constant voltage of 140 V. The maximum hardness of 1000 HV was obtained for a duration of 4.5 h.

6.3 Plasma Radical Nitriding

Lee et al. [4] studied the plasma radical nitriding behavior of SKD 61 tool steel. The treatment was done in the atmosphere of 20 % NH_3 + 80 % H_2 with pressure of 133 Pa voltage of 500 V and at temperature of 450–550 °C for 1–10 h.

It was observed that in this method, no compound layer was formed when the treatment conditions were well-controlled. This is an advantage due to the no requirement for final finishing treatment.

By increasing the time and temperature of treatment, the thickness of diffusion layer was reported to increase up to 150 μm. For the sample treated at 500 °C for 10 h, the formation of compound layer was reported. Therefore, the duration should not exceed 5 h to avoid the formation of compound layer.

The surface hardness of radical nitrided layer was reported to be two times higher than that of the untreated surface. Also, the increment of time and temperature of treatment led to the hardness increasing of the diffusion layer.

The main constituents of diffusion layer were characterized as γ'-Fe_4N and ε-$Fe_{2-3}N$. Again, the amount of these nitrides increased with time and temperature due to the nitrogen diffusion depth. However, at 550 °C, the number of nitrides was

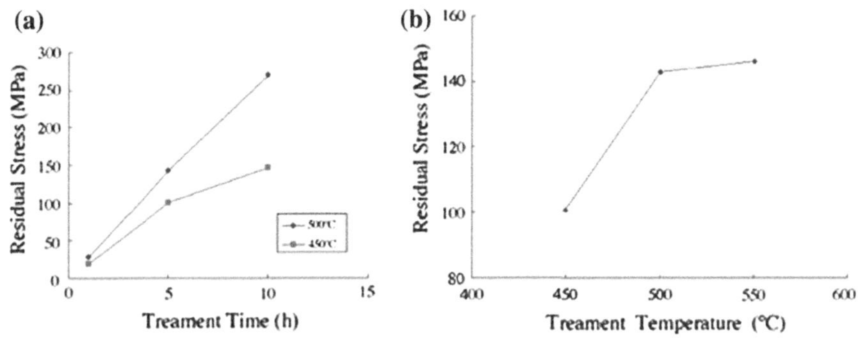

Fig. 6.3 The variation of compressive residual stress as a function of treatment time and temperature [4]

less than 500 °C. The reason was the fact that at this treatment temperature, the solubility of nitrogen in the matrix was high, so that there were not many excess nitrogen atoms left to form precipitates.

The compressive residual stress was also reported by the authors. The increasing of stress was observed with time and temperature as seen in Fig. 6.3. This can be associated with the increase of precipitates with increasing treatment temperature and time. The stress was known to inherently arise from the phase boundaries between precipitates and matrix. Therefore more precipitates led to more compressive residual stresses. The rate of increase of the compressive residual stress in the range of 500–550 °C was much less compared with the range of 450–500 °C. This could be due to the higher solubility of nitrogen in the matrix at the higher nitriding temperature resulting in the decrease of precipitates and the rate of increase of the stress.

The surface roughness for samples nitrided with this method was less than samples nitrided with conventional nitriding methods. This could be due to the external heating of the samples in this method in contrast with conventional DC plasma nitriding in which, the substrate was heated during nitriding caused by sputtering.

References

1. Camps E, Becerril F, Muhl S, Alvarez-Fregoso O, Villagr'an M (2000) Microwave plasma characteristics in steel nitriding process. Thin Solid Films 373:293-298
2. Lin L H, Chen Sh Ch, Wu Ch Z, Hung J M, Ou K L (2011) Microstructure and antibacterial properties of microwave plasma nitrided layers on biomedical stainless steels. Appl Surf Sci 257:7375–7380
3. Abraha P, Yoshikawa Y, Katayama Y (2009) Surface modification of steel surfaces by electron beam excited plasma processing. Vac 83:497–500
4. Lee I, Park I (2007) Microstructures and mechanical properties of surface-hardened layer produced on SKD 61 steel by plasma radical nitriding. Mater Sci Eng A 449–451:890–893

The manufacturer's authorised representative in the EU is Springer Nature Customer Service Centre GmbH, Europaplatz 3, 69115 Heidelberg, Germany. If you have any concerns regarding our products, please contact ProductSafety@springernature.com

Printed and bound by CPI Group (UK) Ltd, Croydon, CR0 4YY
22/07/2026
02174194-0008